MATH JOURNALING ADVENTURES

Logbook ALPHA

EXPLORE THE WORLD
OF CREATIVE MATHEMATICS
WITH DENISE GASKINS

This logbook belongs to:

_____

Year:

_____

Writing is how we think our way into a subject
and make it our own.

Writing enables us to find out what we know,
and what we don't know.

—William Zinsser, Writing to Learn

TABLETOP ACADEMY PRESS

Contents

- Your Mathematical Adventure 4
- Join the Math Rebellion 5
- Character Sheet 6
- Acharya Pingala 7
- Count the Syllable Beats 8
- Warm Up Your Mental Muscles 10
- Jenna Laib's Counting Game 12
- Make a Pattern 14
- Learning 16
- Math Riddles 18
- Five Cards Make 10 20
- Double the Rectangle 22
- Solve Me Mobiles 24
- Menu Math 26
- The Substitution Game: Round 1 28
- Mystery Coins 30
- Malcolm Swan's Partition Products 32
- On This Day in Math 34
- Circles Get Around 37
- Explain a Problem 38
- Multiplication Wheel 40
- Ancient Numbers 42
- Age Puzzles 44
- Stuck 46
- Chocolate Party 48
- Hare and Hounds Game 50
- Square Numbers 1 52
- Julia Robinson Math Festival 54
- Shopping Puzzle 56
- My Secret Rules 58
- Number Yoga 60
- Bowling Game 62
- Linear Crossings 64
- Would You Rather? 66
- Collecting Data 68
- Circle Dance 70
- Juego del Oso 72
- Colored Paper and Metal Disks 74
- Mystery Numbers 76
- How Crazy Can You Make It? 78
- Math Eyes 80
- Basic Nim Game 82
- Two Truths and a Lie 84
- Mini-Biography 86
- Midnight Game 88
- Comparison Puzzles 90
- Reinvent Your Homework 1 92
- Triangles in a Hexagon 95
- Can You Solve It? 96
- Noticing 98

Copyright © 2025 Denise Gaskins. All rights reserved.
Tabletop Academy Press, Boody, IL, USA, TabletopAcademyPress.com

About Math Websites: The Internet overflows with a wealth of math resources, but nothing is forever. If a website disappears, you can run a browser search for the author's name or article title. Or try entering the web address at the Internet Archive Wayback Machine (archive.org).

Contents

- Taxicab Geometry 100
- Which One Doesn't Belong? 102
- Create Your Own WODB 104
- Math Quilt 106
- Math Book Review 108
- Everything Is a Rectangle 110
- Contig Game 112
- Math Translation 114
- Array Puzzle 116
- Career Math 118
- The Answer Is… 120
- Circle Pattern 122
- Number Sums 124
- In the News 126
- Words Help Us Think 128
- Chicken Nuggets 130
- Litton Game 132
- Fraction Wall 134
- Visual Patterns 136
- Create Your Own Pattern 138
- It's Easy 140
- Infinite Series 143
- Math Poetry: Limerick 144
- Sam Vandervelde's Criss-Cross 146
- Silly Definitions 148
- Counting Squares 150
- Explain a Mistake 152
- Square Numbers 2 154
- Living the Dream 156
- Peter Harrison's Cow Puzzles 158
- That's Mean 160
- Explain How 162
- Mental Math Workout 164
- Secret Number Codes 166
- The Mighty Cats 168
- Ancient Chinese Pinwheel 171
- Don Steward's Swap 172
- Six-Word Stories 174
- One Won Game 176
- Math Pickle 178
- Triangular Numbers 180
- Memory 182
- Create a Font 184
- Math Report 186
- Math Poetry: The Square 188
- All Ones 190
- Place Value Mastermind 192
- Math Riddles Redux 194
- Debate with George Pólya 196
- Reinvent Your Homework 2 198
- Captain's Log 200
- Classify the Aliens 202
- Slime Trail Game 204
- Strategic Thinking 206
- Don Steward's Arctan Puzzles 208
- The Substitution Game: Round 2 210
- Rear-View Mirror 212
- Special Thanks 214
- Discover the World of Math 215

Your Mathematical Adventure

Welcome to the team of math adventurers—humans throughout history who have joined together to explore the wonder of numbers, shapes, and patterns.

Most people don't realize that math is a social endeavor.

From the first artist scratching geometric designs in the sand to modern mathematicians expanding the frontiers of knowledge, people have always learned best when we learn together, sharing ideas and building on each other's thoughts.

How to Use This Book

This logbook offers more than 100 puzzles, games, investigations, and conundrums—prompts that lead you to discover new ways of looking at math or to rethink ideas you thought you'd mastered.

For some activities, you may not need the whole two-page spread. When you have extra space, you can:

♦ Try to find another, different way to answer the prompt. Sometimes a new approach leads to greater insight.

♦ Fill the page with a geometric doodle or a pattern of your own design.

♦ Write math rebel answers to your homework problems.

Other times, you may discover you need more space because your ideas flow beyond the confines of this book.

A Natural Cycle of Learning

There are no "right" answers. Instead, each prompt invites you to notice, wonder, and create your own math:

♦ *Notice* means to open your eyes and pay attention to details, examining all aspects of the situation, seeing beyond just the surface level. Write down everything you observe.

♦ *Wonder* means to respond to the things you notice, searching out relationships and connections to other concepts, diving deeper into the sea of ideas. Write down any questions you think of. Don't worry about answers, just brainstorm.

♦ *Create* means to process the things you notice and wonder, shaping them through your own perspective on the world, to make something new.

You might create an explanation, a story or poem, a drawing, a new question to investigate (mathematicians love finding new questions!), or whatever fits your own unique inspiration.

Like other math explorers through the ages, you may find this process easier and more enjoyable if you share it with friends. In real math, working together is never "cheating." It's the natural way for humans to learn and grow.

May you always enjoy the adventure!

Join the Math Rebellion

Math rebels write any true answer *except* what the textbook expects.

For example, if the textbook answer is 57, a rebel might write:

$$100 - 43$$

$$\text{Or } {}^{120}/_2 + (-3)$$

Or *"The total number of mushrooms in the basket, if three hobbits each picked nineteen 'shrooms (not counting the ones they ate)."*

Math rebels can make the answer as crazy as they like. Have fun!

Understand the Math

When you get a math worksheet or homework page, don't start working straight away. First, review the page to see if the problems look familiar. Do you know what the teacher or textbook wants you to do?

Math rebels always care about the truth. So first, learn what the problem means and how to figure it out. After you know how to solve the problem, then you can start working on your creative answer.

Choose Your Battleground

Fighting for intellectual freedom takes energy. Are you going to mess with just a few of the problems? Or will you turn the entire lesson into a protest statement?

Math rebels know the importance of justification. So be ready to defend whatever you write.

Another Way to Play

Instead of doing a long page of math homework problems on a single topic, follow the natural cycle of learning.

Work the first few problems, until you notice something that makes you wonder. Then spend the rest of your math time investigating that question.

You may not discover the answer to your question, but the adventure of exploring how math works will build a deeper understanding of your topic than a whole page of homework problems done by rote.

Live by the Two Rules

There are only two important rules in mathematics:

♦ You are allowed to write anything that makes sense.

♦ You are not allowed to write anything that doesn't make sense.

Anything else is just advice. Follow it if you wish, or blaze your own path.

Fight for Truth, Justice, and Creative Reasoning

Character Sheet

Fill this page with information and data about yourself. For example, you might include your height, age, the number of people or pets in your family, your birthday, favorite number or shape, history ("5 years ago I ...") or future ("In 3 years, I will ...").

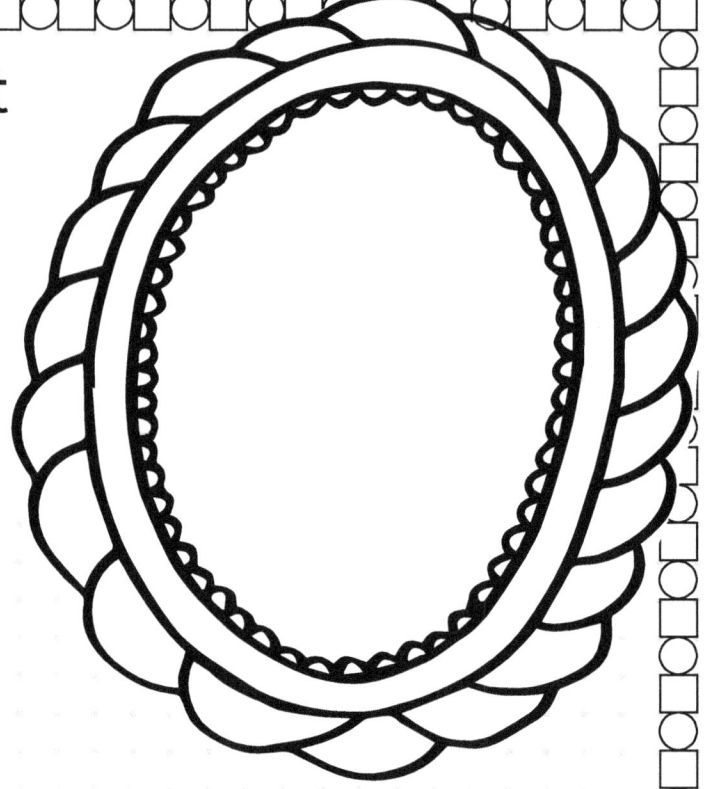

CHALLENGE: Write the facts Math Rebel style, using an expression in place of each number. For example: "I have $8/4$ dogs."

Acharya Pingala

Pingala lived in India in the 3rd or 2nd century BC. His title "Acharya" means "honored expert teacher."

A poet as well as a mathematician, Pingala wrote the *Chandahśāstra* (also called *Pingala-sutras*). His book contains the earliest known description of a binary numbering system.

Based on his study of Sanskrit poetry, Pingala wrote about combinatorics, the binomial theorem, and the "Fibonacci" number pattern (many centuries before Fibonacci was born). Pingala called that number series *mātrāmeru*. He found the mātrāmeru by studying poetry.

He is sometimes credited with the first use of zero as a number, which he called *śūnya* (Sanskrit for "emptiness" or "nothing").

Try your hand at Pingala's poetic math with the "Count the Syllable Beats" puzzle on page 8.

*Acharya Pingala,
Math Team Alpha's Leader Emeritus*

What Is "Alpha"?

Mathematicians use many letters of the Greek alphabet to stand for different constants, variables, functions, etc.

The first three letters are *alpha*, *beta*, and *gamma*. We often use their lowercase forms α, β, and γ as variables to label unknown angles in a diagram.

For example, see "Don Steward's Arctan Puzzles" on page 208.

Count the Syllable Beats

In English poetry, meter depends on accented syllables. In Sanskrit poetry, meter describes the pattern of short and long syllables.

   A short syllable takes one beat, like a quarter-note in music, while a long syllable takes two beats like a half-note.

   Acharya Pingala wondered, "If a poet has a certain number of beats to fill using short and long syllables, how many different ways can he arrange them?"

   One way to show the beat patterns is by using two lengths of blocks: a square for the short syllable and a rectangle for the long one.

   Explore how the number of possibilities change as the lines of a poem grow longer. What do you notice about the syllable beat patterns? What do you wonder?

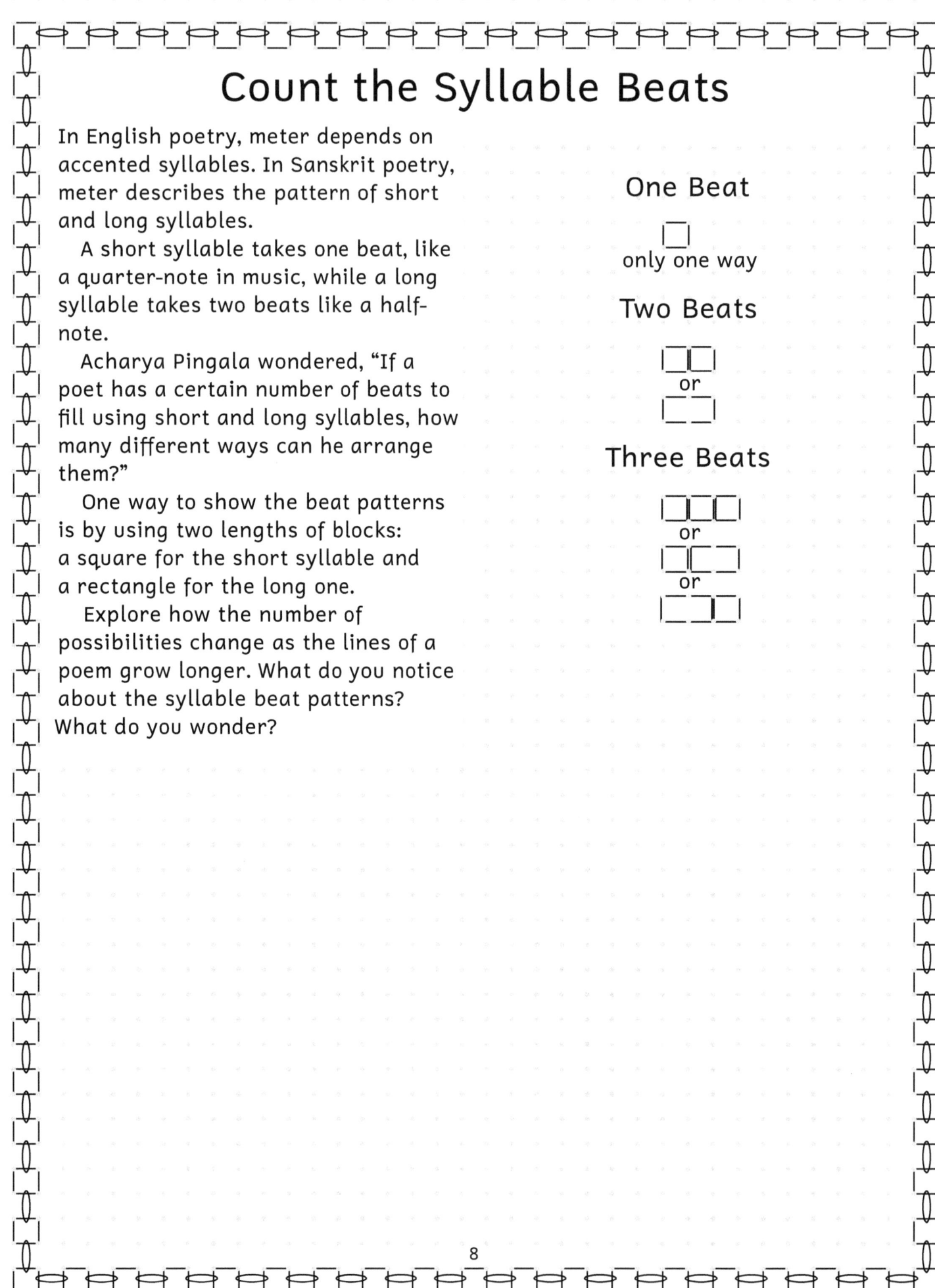

Warm Up Your Mental Muscles

Practice your Math Rebel skills. Pick any number, and see how many different ways you can write it. Start with a simple expression like addition or multiplication. Write an equal sign and another expression. Think about ways you can modify each expression to create a new one, and keep going until you fill the page or run out of ideas. What kind of fancy math will you create?

Jenna Laib's Counting Game

(two players or small group)

Choose a whole number, fraction, decimal, or negative number to skip-count by. Choose a starting place and a target number. For example, "Count by threes from 11 to 47."

Each player in turn chooses to write one, two, or three skips of the counting-by number. For example, the first player may write "11, 14." The second player might add "17," and then the next move could be "20, 23, 26." And so on.

The player who reaches or passes the target number wins.

Count by _____

Start at _____

Target = _____

Extension: Mathematicians love to tweak games just to see what happens. How would you modify this game? Test out your new rules with a friend.

Make a Pattern

Create a pattern of numbers, shapes, or colors. Keep it going as long as you can. Will your pattern go all the way around the page? Or begin at the center and spiral outward? Or make any other design you like.

Need ideas? Do a search for "frieze patterns."

Learning

Finish the prompt sentence. And then keep writing until you run out of room. Don't overthink it, just write. Keep your pencil moving. If you can't think of what to write, copy your previous sentence over and over until your mind comes up with something new to say.

The best way to learn math is...

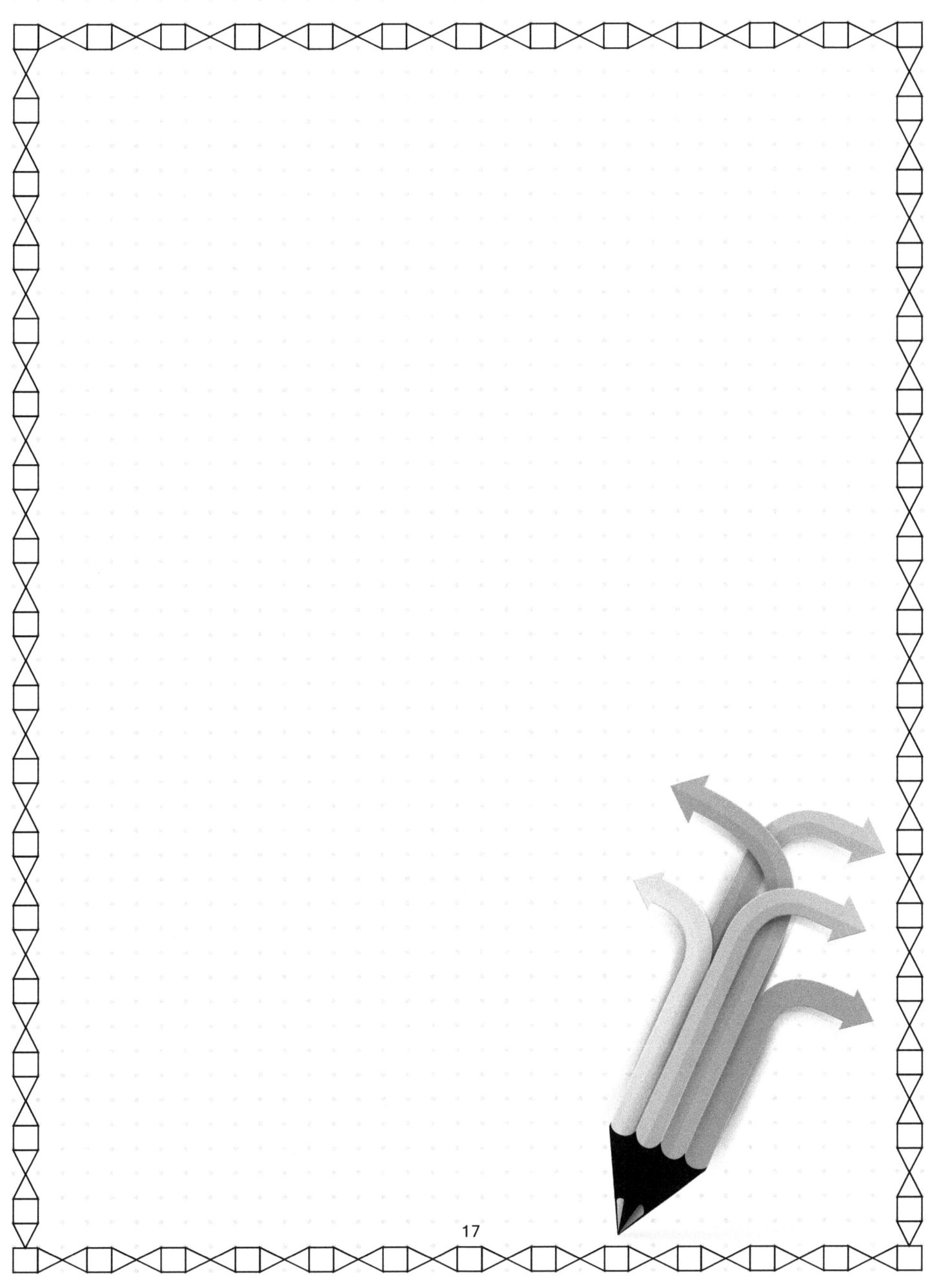

Math Riddles

(any number of players)

Choose a secret number the other players will try to guess. Write a "What Number Am I?" riddle.

For example, "I am odd and prime. I'm a two-digit number less than 30. The sum of my digits is 4. What number am I?"

Give at least three clues for your mystery number. No other number should match all the clues.

EXTENSIONS: For more sample riddles, go to solveme.edc.org and click "Who Am I?"

Five Cards Make 10

Turn up five ordinary playing cards. (Ace = 1, Jack = 11, Queen = 12, King = 13, Joker = 0, and number cards count at face value.) Use one or more of these numbers to write a math expression that equals 10. You can use any math operations you know.

   Then find another way to make 10. How many different ways can you say "10" with the numbers on your cards?

CHALLENGE: What other numbers can your cards make? Are any numbers impossible? How do you know?

Double the Rectangle

The length and width of a rectangle are both increased by 2 units. The new rectangle has twice the area of the original. What might be the original length and width? (There are several possible answers.)

Solve Me Mobiles

Go to solveme.edc.org and click "Mobiles." Find a puzzle you like. Copy it in your journal. Explain how you figured it out.

Try as many puzzles as you wish. Then make up a mobile puzzle of your own.

Menu Math

Create a menu for an imaginary restaurant. Include main dishes, side dish items, drinks, and desserts. Write a story for your restaurant. What math questions might you ask about your story?

The Substitution Game: Round 1

(solitaire or small group)

Write a simple equation at the top of your paper. On your turn, write on a new line. Copy the equation from the line above, except replace any single number with an equivalent expression. For example:

$$2 + 5 = 7$$

$$2 + 8 - 3 = 7$$

$$2 + 8 - 3 = 14 \div 2$$

$$2 + 8 - 3 = (100 - 86) \div 2, \text{ etc.}$$

Before you leave, copy your final equation at the top of page 210.

Mystery Coins

I have eight coins in my pocket. How much money might I have? Is there any amount that's impossible? Why?

Malcolm Swan's Partition Products

Pick a number. Break it up into as many pieces (called *partitions*) as you want. Now multiply all those pieces together. Try a different set of partitions. What's the biggest product you can make?

For example, suppose your original number was 25. You might partition it into 1 + 1 + 1 + ... + 1, but that set has a rather disappointing product. (Why?) Or you could try 21 + 4, which multiply to make 84. Or 10 + 10 + 5, which makes a nice, big product of 500. Can you do better than that?

CHALLENGE: What if your partitions don't have to be whole numbers? For example, what if you split 25 into 2.25 + 4 + 6.25 + 12.5?

On This Day in Math

Go to pballew.blogspot.com and find today's "On This Day in Math" post. Choose one of the events or people mentioned and summarize that information here. Do a web search to see what else you can find out about that person or event.

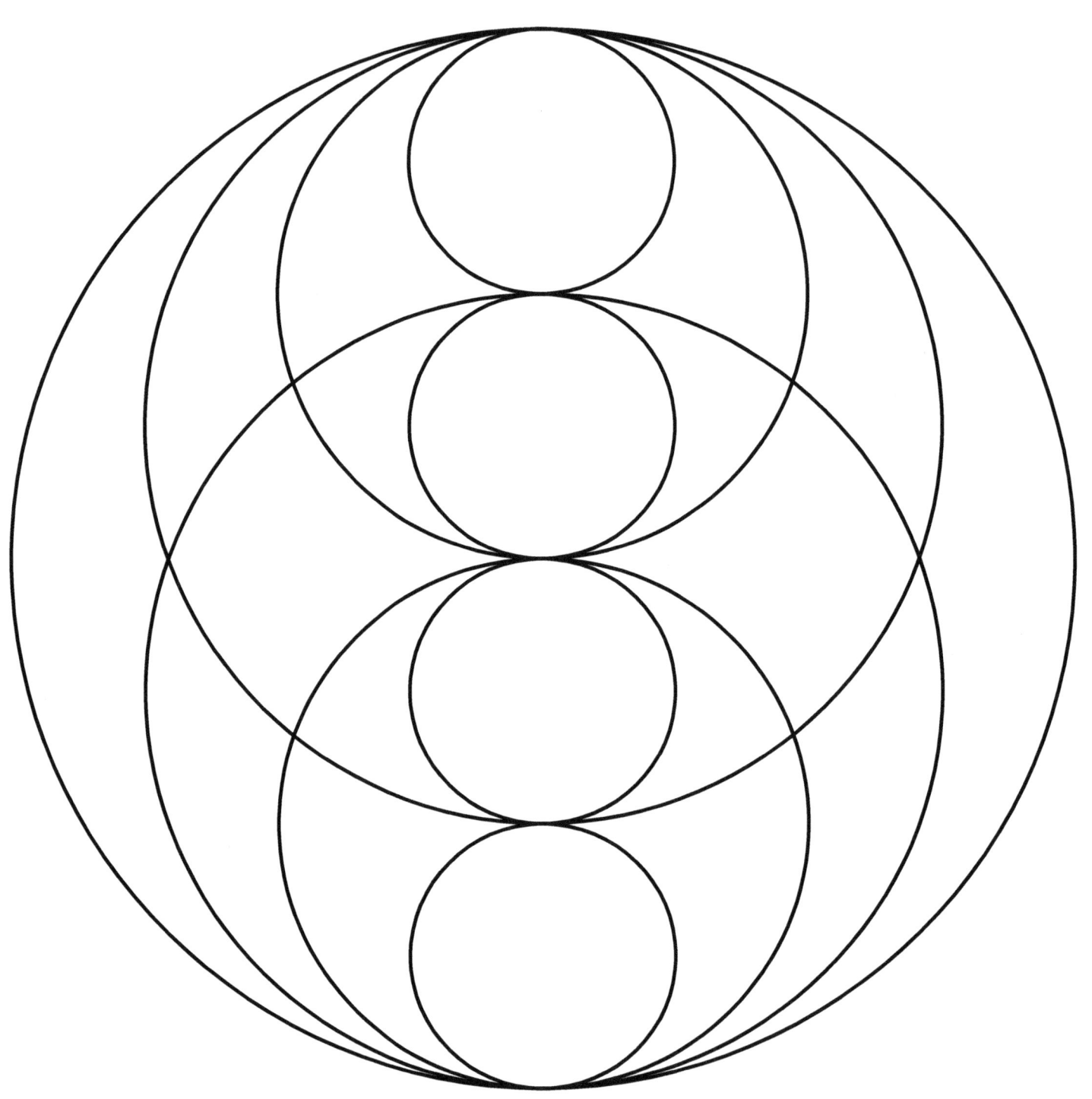

Circles Get Around

Look at the math-art design. What can you say about the shapes or angles? Make a list of the things you notice. What do you wonder? Color the picture, or fill each section with a pattern.

OPTIONAL: Create a related math-art design of your own.

Explain a Problem

Copy a story problem from your math book, but don't include the numbers. Can you explain how to solve it without using any numbers?

For example: "Joseph knows the price of a box of candy and the price of a certain book. How can he figure out how much money he will have left after buying them both?"

One possible solution: "Joseph can add the prices together. If there's a sales tax, he also has to add in that percentage of the sum. This will give him the cost of his purchase. Then he can subtract this cost from the total amount of money he has, to find out how much will be left."

EXTENSION: If you enjoy the challenge of solving problems without numbers, you might like Farrar Williams's book *Numberless Math Problems: A Modern Update of S.Y. Gillian's Classic Problems Without Figures.*

Multiplication Wheel

Draw a circle and write any numbers or algebraic expressions around the inside, arranged like the numbers on a clock. Draw spokes out from the center, dividing your wheel into one sector (slice-of-pizza shape) for each number.

Extend the spokes out to reach a second, larger circle. In this circle, write the double of each sector number. On the next time around, write the triple of each sector number, etc.

How large will your multiplication wheel grow?

Ancient Numbers

Look up how to write numbers with Roman or Mayan numerals. Can you write your birth year in that style?

How would you do math calculations with numerals like these? Make an infographic using pictures and words to explain the counting system.

Age Puzzles

Jewel's father is 3 times as old as Jewel. In 10 years, he will be twice her age. How old is Jewel? Make up some age puzzles of your own.

Stuck

Finish the prompt sentence. And then keep writing until you run out of room. Don't overthink it, just write. Keep your pencil moving. If you can't think of what to write, copy your previous sentence over and over until your mind comes up with something new to say.

When I get stuck on a math problem...

Chocolate Party

You are hosting a party for your friends, and they all love chocolate. So you bought six super-king-size bars for everyone to share. You put three bars on the largest table in your party room, two bars on the mid-size table, and one bar on the smallest table.

As each friend enters the room, they choose a chair that will give them the largest piece of chocolate when all the bars at that table are shared evenly. (But of course, the chocolate won't be shared out until all guests arrive. And guests are not allowed to move once they're seated.)

How many guest have you invited? In what order will the chairs around your tables be filled? Is there a perfect "chocolate party" number, which gives each guest the same-size piece of candy? Can there be more than one such number?

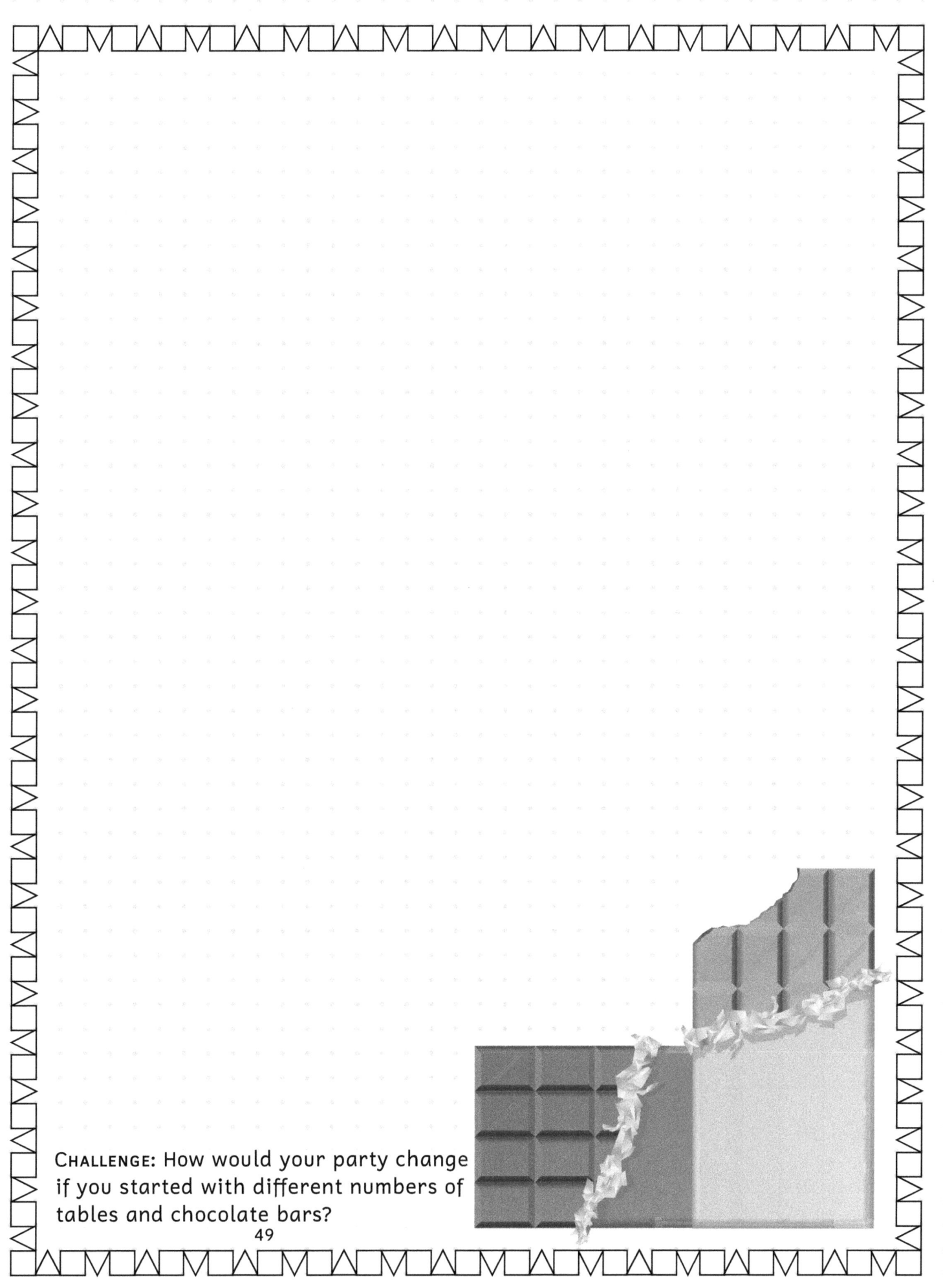

CHALLENGE: How would your party change if you started with different numbers of tables and chocolate bars?

Hare and Hounds Game

(two players)

Players each need identifiable tokens (or crumpled bits of colored paper). One player controls three hounds, which start at the top (North) of the gameboard. The other player is the hare, which starts at the bottom.

   The hounds may move from one circle to the next, down or sideways along the lines, but never back toward the North. The hare may move one space in any direction, but may not jump over a hound.

   The hounds win if they trap the hare. The hare wins if it gets past the hounds.

EXTENSION: Play several games, keeping track of who goes first and who wins. Do you think this is a fair game, or does one player have an advantage?

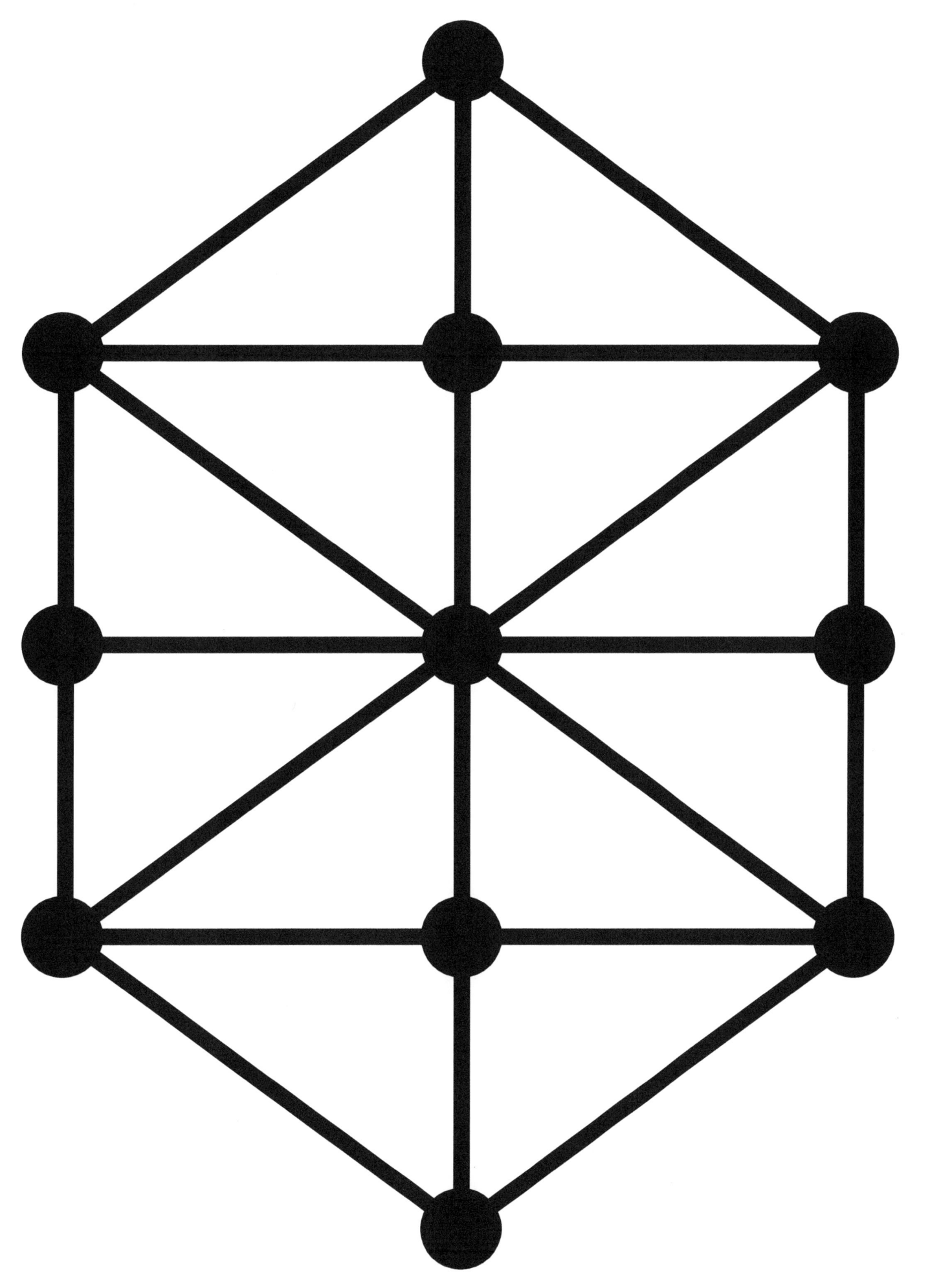

Square Numbers 1

Draw several different squares of different sizes, counting along the grid lines to make each side the same length.

When you draw a square with sides of S (any number) grid units, the area (A) will be S squared:

$$A = S^2$$

And the side is the square root of the area:

$$S = \sqrt{A}$$

Label the sides of your squares and count the square units for their areas. Make a list of the square numbers you find (the areas) and their square roots (the sides).

What do you notice about the numbers or the sizes of the squares? Can you think of any questions to ask?

Julia Robinson Math Festival

Go to jrmf.org/puzzle. Choose a game or puzzle. Try it several times. Find different ways to play with the challenge. Write about your discoveries.

Shopping Puzzle

The Horticulture Club has a plant sale: Flowering plants are $_____ each, and hanging plants are $_____ each. [Choose any two numbers.]

Sasha has $100 to spend. What might she buy?

Make up your own shopping puzzle.

My Secret Rules

(two or more players)

Choose a secret number rule for each of the circles in the Venn diagram. For example: even numbers, prime numbers, and multiples of 5. The other players take turns guessing a number, and you place it in the circle for the rule it matches.

Numbers that match more than one rule go in the section where the circles overlap. If the number doesn't match any of your rules, it goes outside all the circles.

Play until the others can name all of your secret rules.

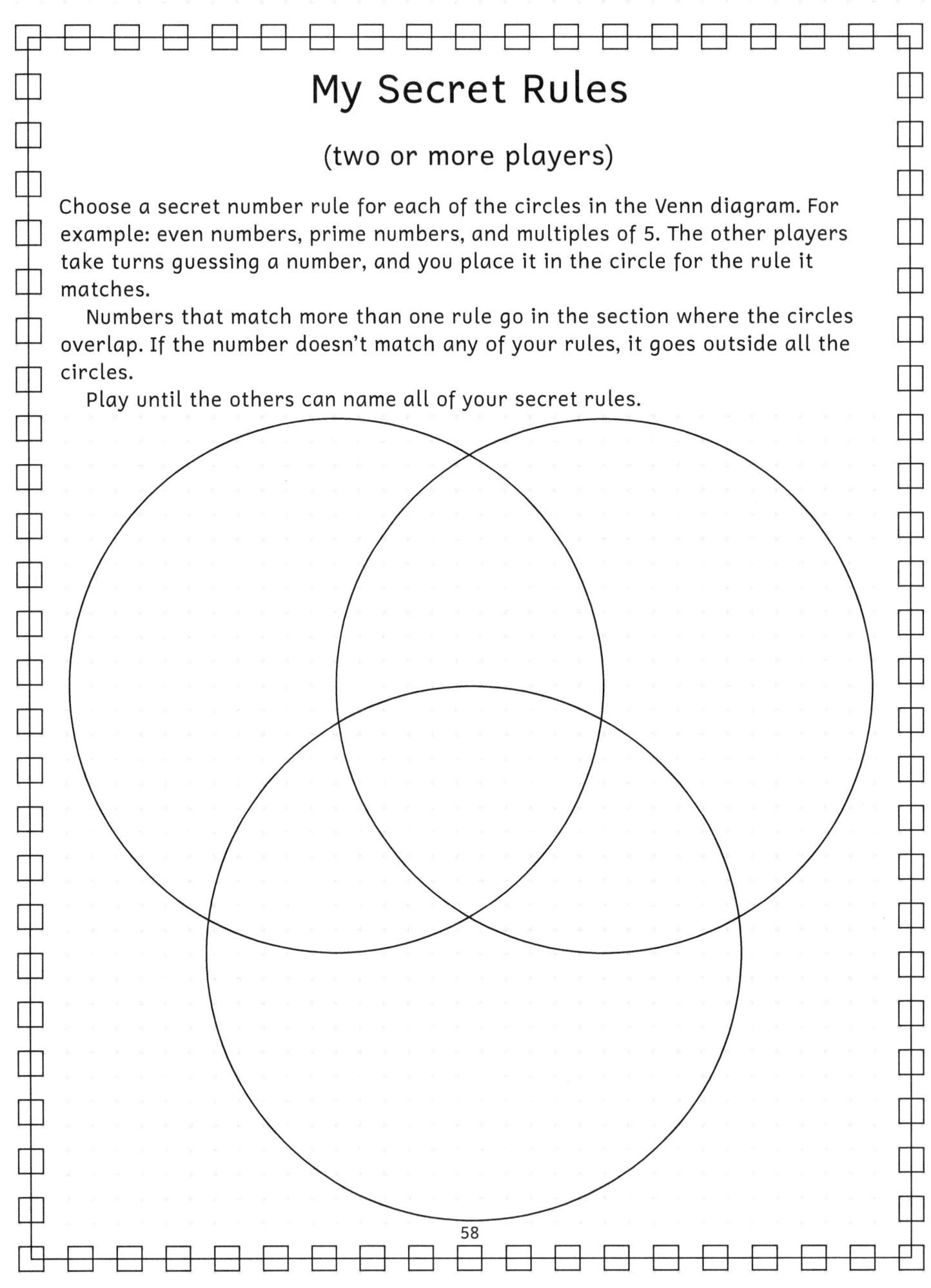

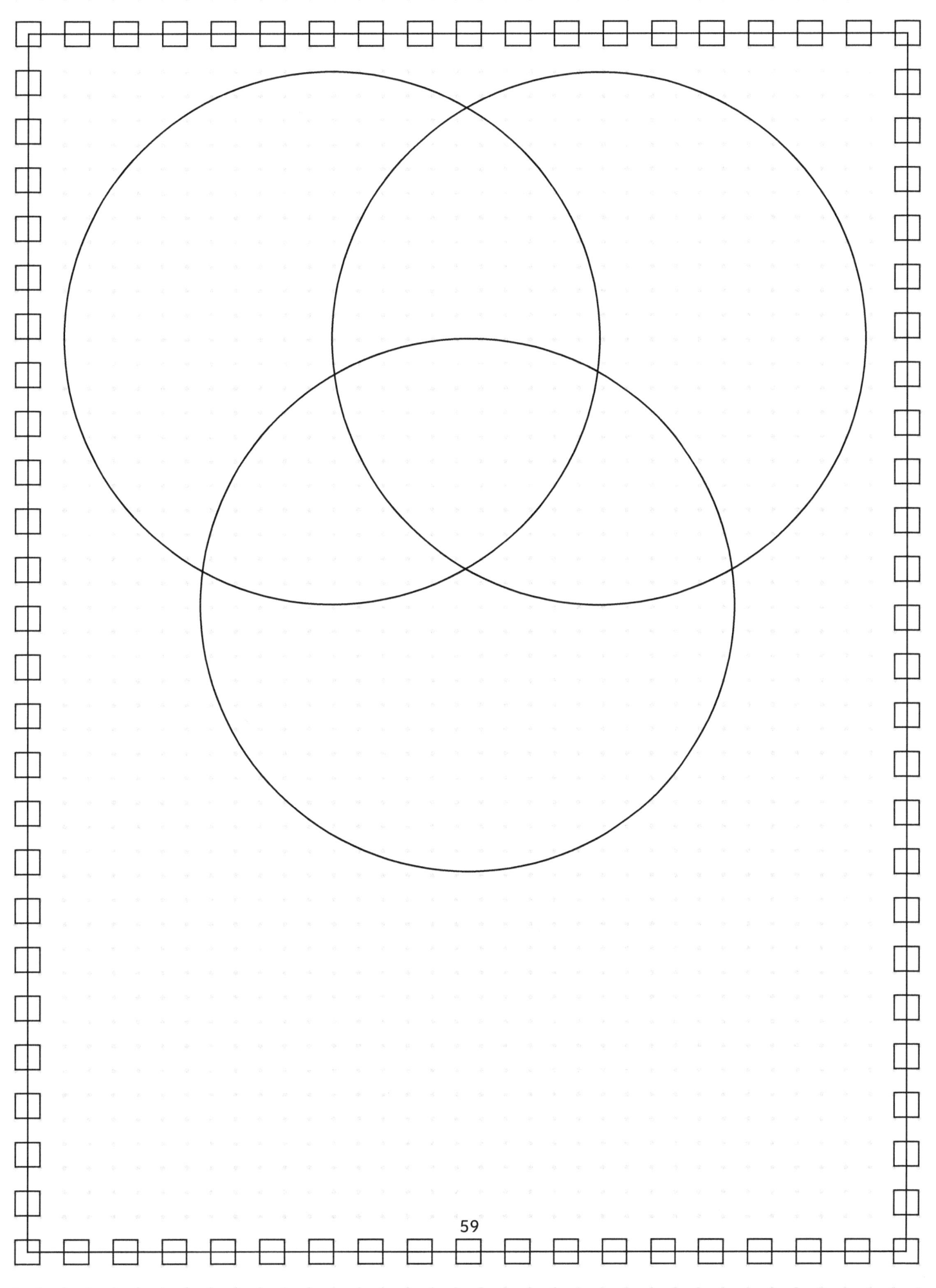

Number Yoga

Use the digits 1, 2, 2, and 6. You must have all four digits in each expression, with each digit appearing only once. You may use any math operation you know: +, −, ×, ÷, brackets, etc.

For example: 1 + (2 × 2 × 6) = 25, and 62 − 12 = 50.

Can you calculate all the numbers 1–10? 11–20? What other numbers can you make? Are any numbers impossible to make with your four digits? (Or perhaps they are possible, but not with the math you've learned so far?)

Bowling Game

(solitaire)

Draw circles in a bowling-pin pattern. Write the numbers 1–10 in the circles. (Bowling pins are arranged in a triangle and numbered from the shortest row to the longest.)

    Roll two dice and cross out any combination of circles that exactly matches that sum. Those are the pins you knocked down. Roll again, trying to hit more pins. Keep going until you roll *a gutter ball*—a set of dice that you can't match.

    If all the numbers left are 6 or less, you may choose to roll only one die. Can you knock down all the pins?

Linear Crossings

Draw straight lines. Each line must cross all the others, but they may not cross at the same point. How many lines can you draw?

Try the experiment again. Can you get more lines crossing this time?

Do you think there is a maximum number of straight lines that all intersect each other? If they are allowed to cross at the same point, is there a maximum?

Would You Rather?

Go to wouldyourathermath.com and click your grade range. Find a WYR question you like, and copy it in your journal. Explain your answer: Which would you prefer, and why?

Collecting Data

Choose something you are curious about that can be measured. For example: the outdoor temperature or rainfall, or how many jumping jacks you can do in a row, or how much time you spend on social media.

  Measure it every day for at least a week. Write a list of things you notice about the numbers that you find. Make a chart or graph to visualize your data. Remember to label your chart or graph, so people can tell exactly what you measured.

Older students: Try several different styles of graph. Which one do you like best for this type of information?

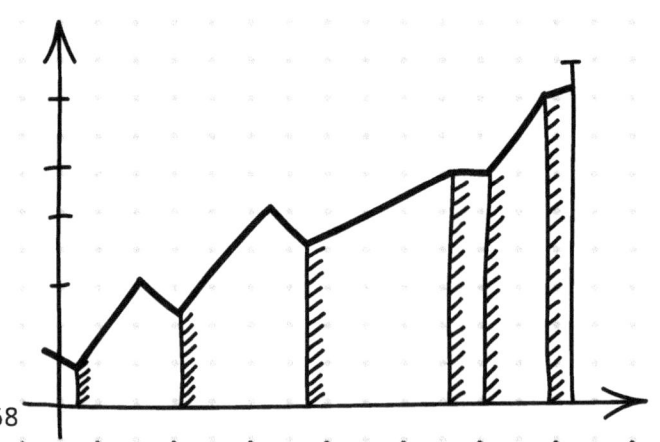

Circle Dance

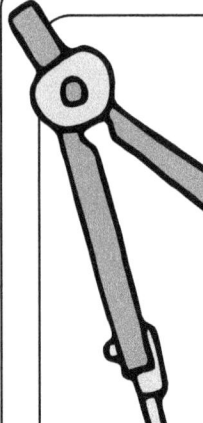

Drafting compasses can be tricky to control. Practice using a compass to draw circles on your page. Make as many as you like, in different sizes.

Or trace around circular shapes like cups or cans. Let some of the circles overlap each other or get wide and run off the page. What do you notice about your circles?

Color your design, or fill each section with a pattern.

Juego del Oso

(two or more players)

Begin by drawing a 3 × 3 or larger tic-tac-toe grid. On your turn, mark an O or S in any square. Both players may use both letters as they wish.

If you complete the word "OSO" in any direction—vertically, horizontally, or diagonally (with no space between letters)—mark it out and add one tally mark to your score. Then take a free turn, writing another letter.

The same letter may be used in more than one OSO, each going in a different direction. You may complete several OSOs in a row, but always end your turn by marking one more letter.

When the board is full, or no more OSOs are possible, then whoever gained the most points wins the game.

Colored Paper and Metal Disks

What is money? Could you explain it to an alien from a planet that doesn't have money?

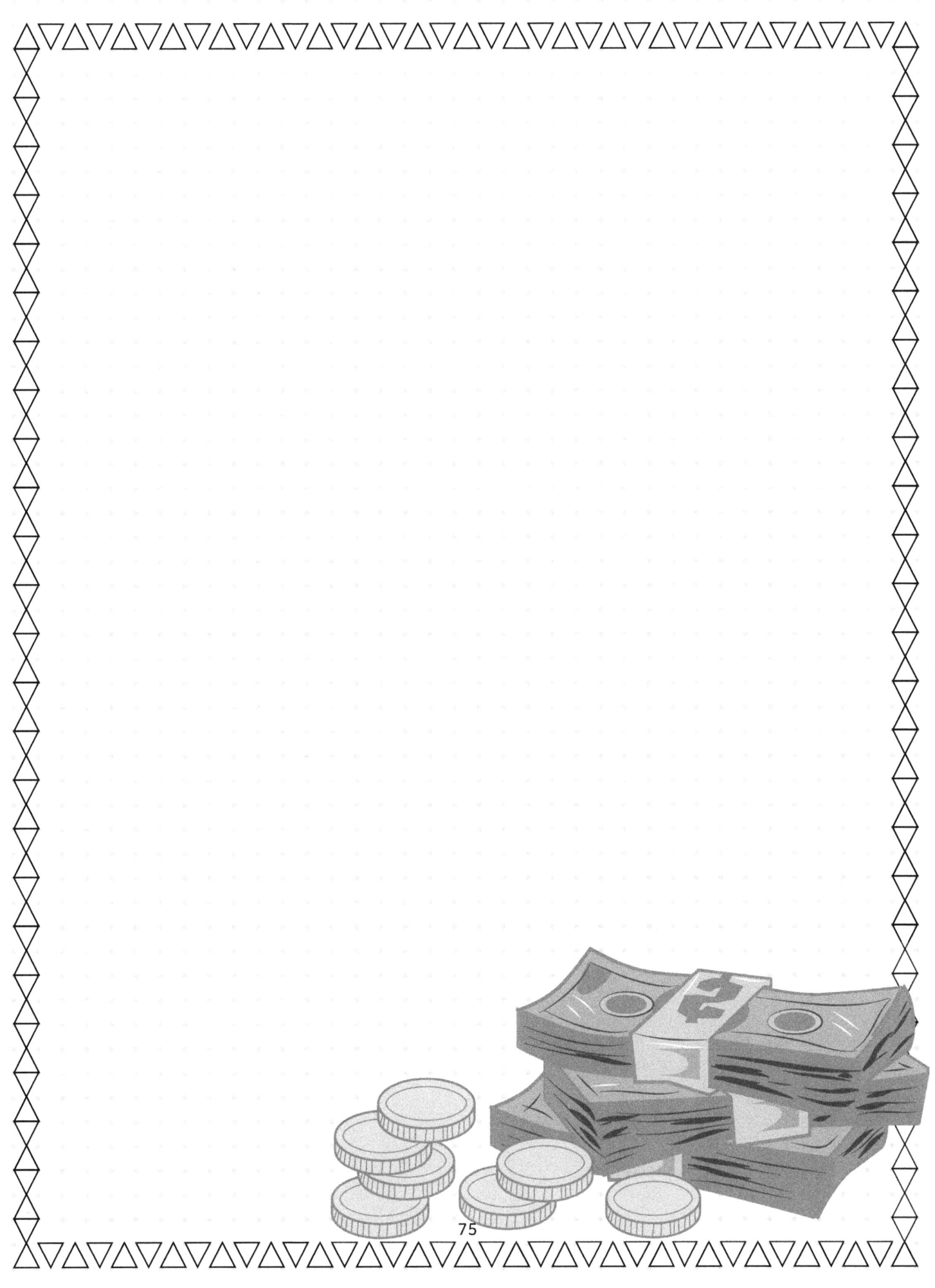

Mystery Numbers

Write equations with missing mystery numbers, like

3 + □ = 57

or

(5 × 5) + 4 = 16 + □ − 3

Or use algebra:

79 = (n × 11) + 2

Or be silly:

giraffe ÷ 4 = 37 − 12

Can you solve your mystery number equations? Or trade puzzles with a friend.

How Crazy Can You Make It?

Choose any target number you like. Fill each shape with an expression that equals the target. Can you make some cool, creative math?

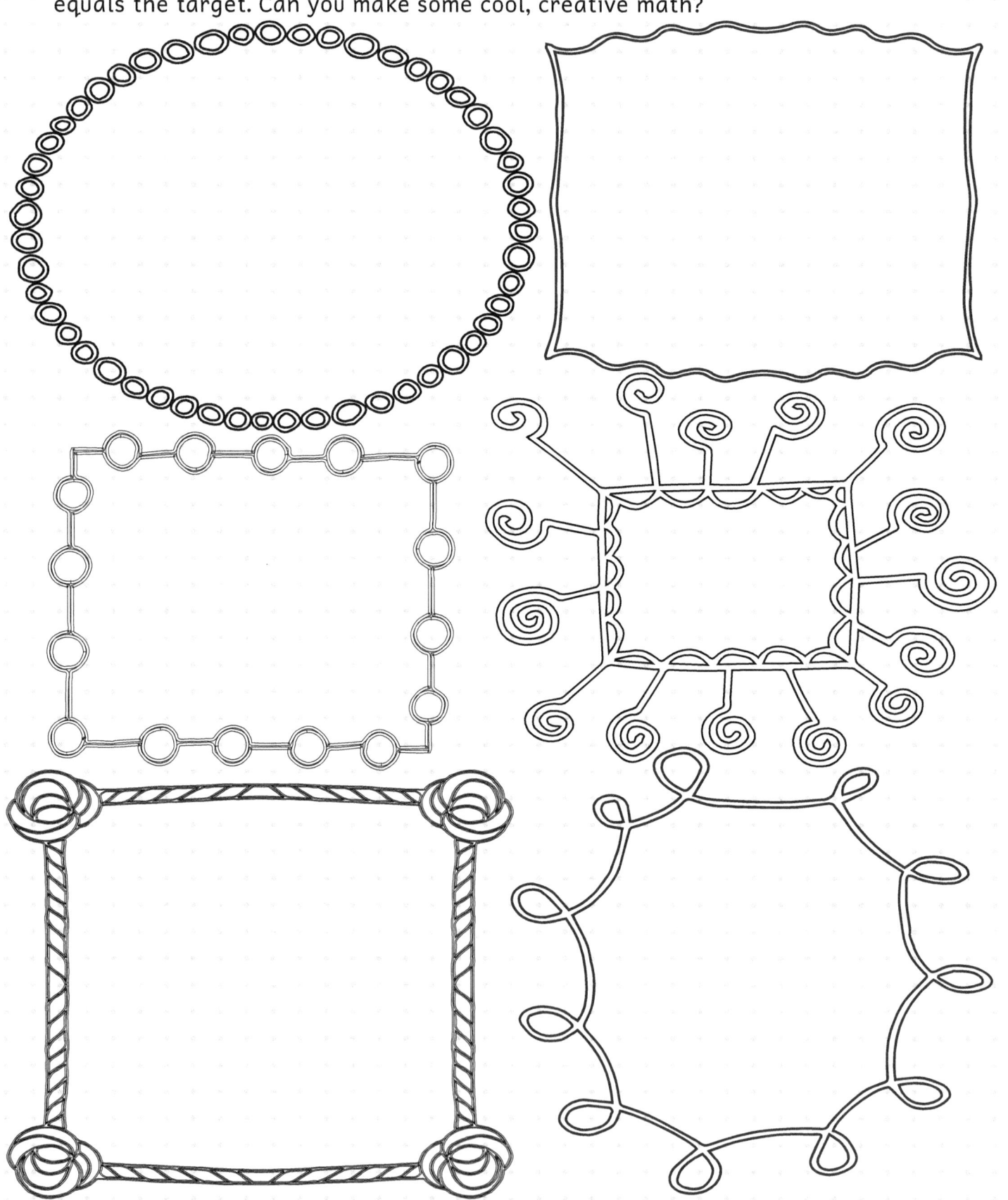

Math Eyes

Imagine you have special glasses that let you see everything through the lens of math. Choose an item from your room or something in your house. Don't tell what it is, but describe it using as much math as you can. Think about measurements, shape, position, designs or patterns, motion, etc.

Challenge: Read your description to someone. Can they identify your object?

Basic Nim Game

(two players)

Draw 10–15 circles (called "stones"). On your turn, mark out one or two of the stones, removing them from play. Whoever marks the last stone wins the game.

In future games, whoever lost the previous round chooses whether to go first or second.

Misère variation: Whoever marks the last stone loses.

Two Truths and a Lie

Pick a topic you have learned in math. Write two correct statements and one false statement. Trade with a friend. Can you find each other's fibs?

Mini-Biography

Write about a Black mathematician who is NOT Benjamin Banneker. Or write about a Latino or Indigenous mathematician.

Look at mathigon.org/timeline or arbitrarilyclose.com/mathematician-project or mathshistory.st-andrews.ac.uk for ideas.

Midnight Game

(two or more players)

Players agree on a time of day for the clock to begin, anywhere from 1:00 AM to 12:00 noon. Write the time at the top of your paper.

On your turn, you may move either the hour hand or the minute hand ahead by one number. (That is, add one to the hour, or 5 to the minutes.) Write the new time on the next line.

The player who gets to write "12:00 AM" (midnight) wins the game.

EXTENSION: Mathematicians love to tweak games just to see what happens. How would you modify this game? Test out your new rules with a friend.

Comparison Puzzles

A melon weighs as much as 5 premium apples, or as much as 15 plums. How do plums and apples compare?

If a grocery bag is strong enough for 2 melons, what other combinations of fruit can it hold?

Make up a measurement-comparison question of your own.

Reinvent Your Homework 1

Find a page of calculations in your math book, or download a worksheet online. Choose two or three of the questions. Write a story problem to match each calculation.

For example, for the calculation $3/4 \times 8$, you might imagine a recipe that takes $3/4$ cup of flour. But you are planning a party and need to make eight times that amount. How much flour will you need in all?

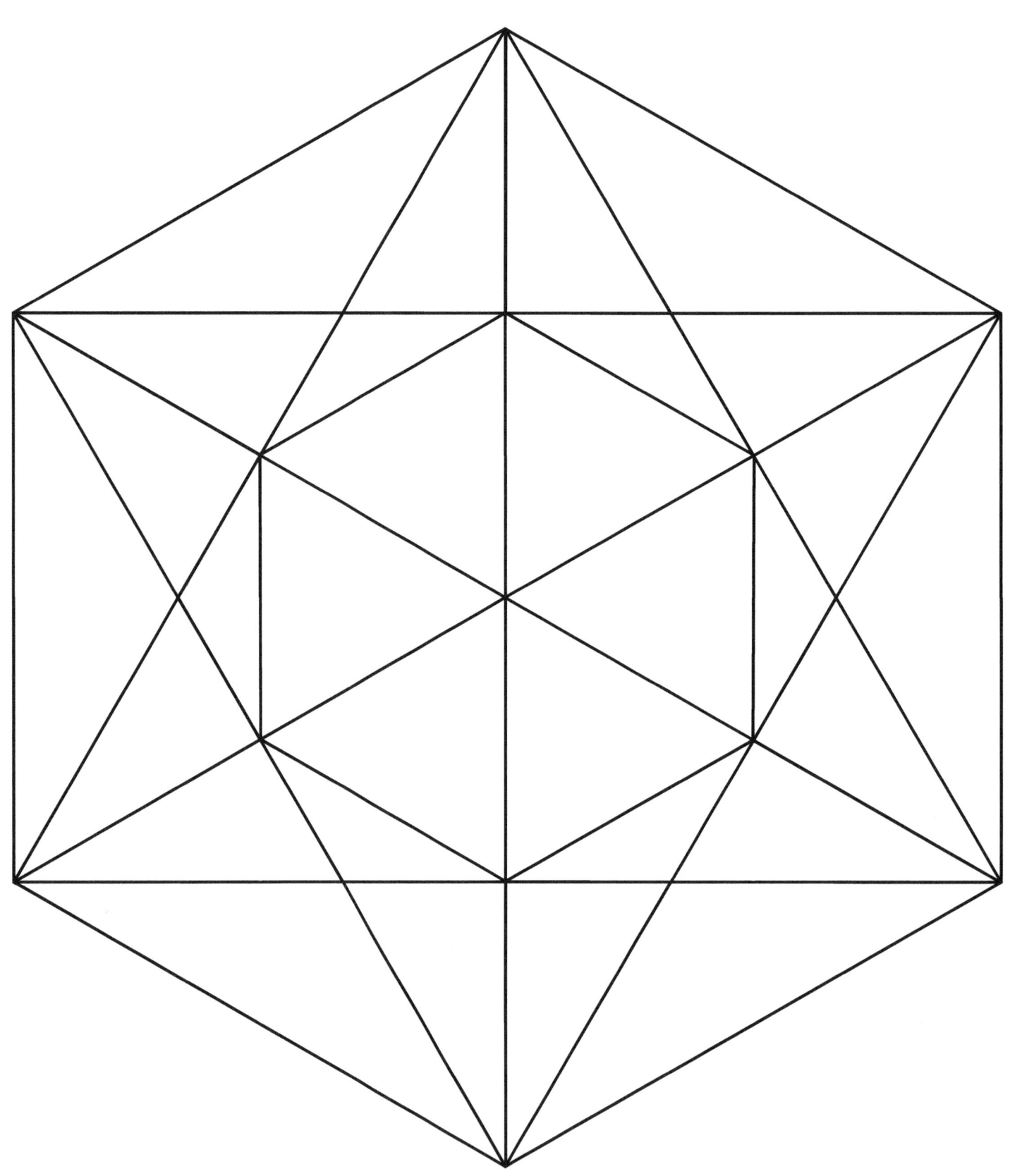

Triangles in a Hexagon

Look at the math-art design. What can you say about the shapes or angles? Make a list of the things you notice. What do you wonder? Color the picture, or fill each section with a pattern.

OPTIONAL: Create a related math-art design of your own.

Can You Solve It?

Go to expii.com/solve or search for the MATHCOUNTS Problem of the Week. Find a math problem you like. Copy it in your journal.

Or copy a problem from your math book.

Explain how to solve your problem. Try to make your explanation clear enough for a younger sibling, cousin, or friend to understand.

Challenge: Can you show more than one way to figure it out?

Noticing

When was the last time you found something mathematical in actual life? Describe what you saw and tell how it relates to math.

Or if you can't think of anything, then look at the room around you and describe whatever math you see. Think about numbers, shapes, lines, curves, and patterns.

Taxicab Geometry

Draw a rectangular grid to represent your city, with squares big enough to write in. Choose any square for your starting position and label it 0 (zero).

Move from square to square horizontally or vertically, like a taxicab on roads that go east-west or north-south. In taxicab geometry, you never, ever move diagonally. In each square, write the least number of moves (shortest path) to get there from zero.

What patterns do you see? Can you think of any questions to ask?

Which One Doesn't Belong?

Go to talkingmathwithkids.com/wodb and find a puzzle you like. Make a sketch of it in your journal. Write out your answer(s) to the puzzle, using arrows or diagrams as needed.

Create Your Own WODB

Choose three attributes an object might have. For example, you might pick pointy, blue, and striped. Then draw four pictures. One picture should have all three attributes, and the other pictures should each be missing a different one. Ask a friend to say which one doesn't belong and why.

Math Quilt

For each piece of the quilt, color a different fraction of the square. For example, one square might represent ½, but drawn creatively (not just a line down the middle).

Or color the same fraction in every square, but make each one look different. What do you notice about your quilt? Does it make you wonder?

Math Book Review

Find a math book at your library. Read it and write a summary or review.

If your library uses the Dewey Decimal System, mathematical nonfiction is on the 510–519 shelf, and logic puzzle books are at 793.74. You can ask the librarian to help you find a mathematical picture book or fiction story or the biography of a mathematician. Or do a search for "CSMP storybooks by grade level."

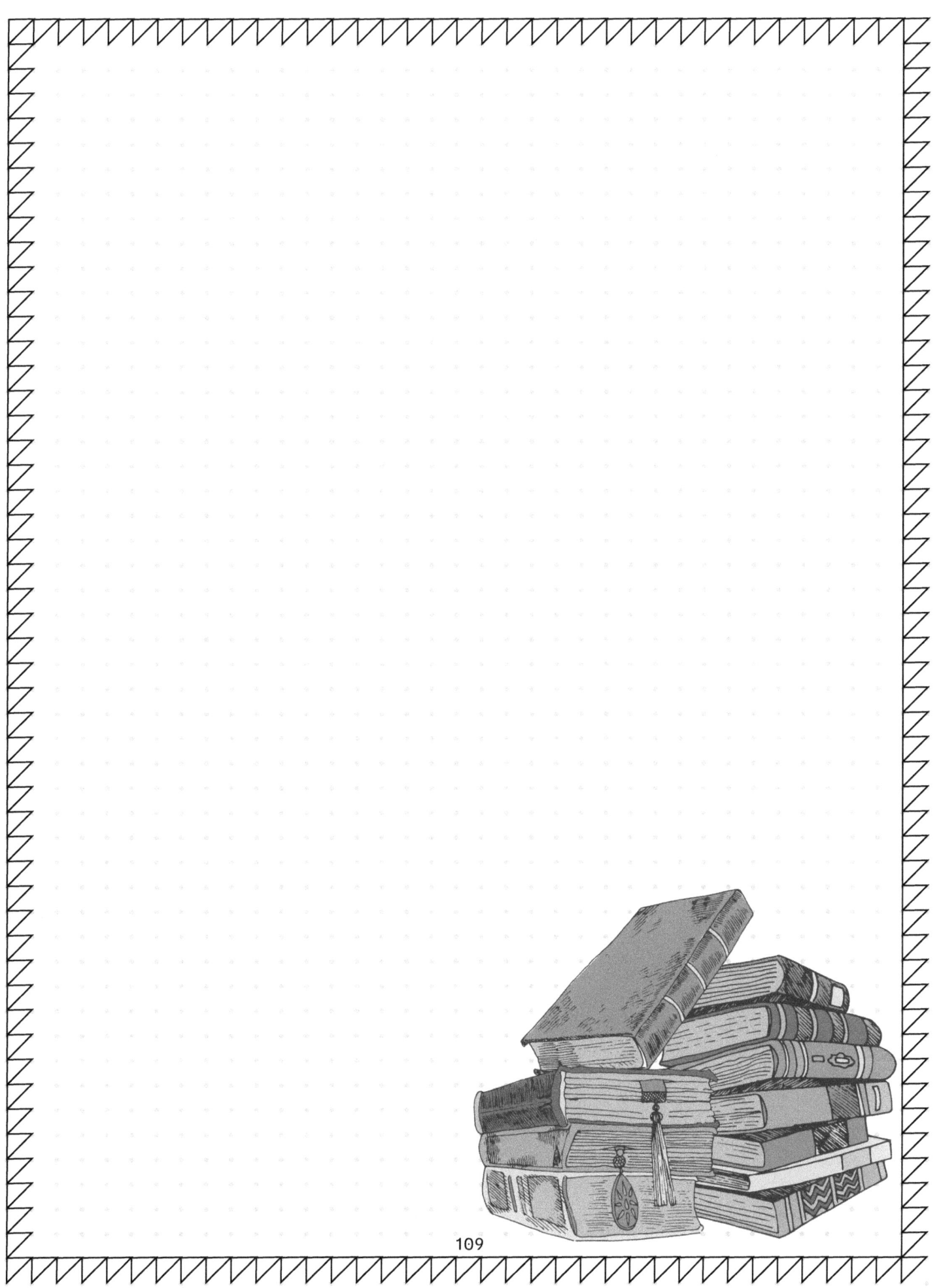

Everything Is a Rectangle

Draw any quadrilateral (four-sided shape) on your page, even an unusual shape like a kite or an arrowhead. How can you convert it into a rectangle with the same area?

    For example, can you imagine cutting off one part and pasting it in a different place? What changes as you move the pieces? What stays the same?

CHALLENGE: Can you convert shapes that are not quadrilaterals?

Contig Game

(two or more players)

On your turn, roll three dice. Use those numbers and basic arithmetic (+, −, ×, ÷) to calculate the number in any unmarked square on the board. Mark your square with a large X. Say out loud how you made the number.

Score 1 point for the square you marked, plus 1 point for each already marked square *contiguous* to your number's square—that is, touching any side or corner. The maximum score for any turn is 9 points. If all the numbers you can make have already been marked, you score a zero—but if anyone else finds a valid calculation using your dice, that player may mark the square, and "steal" those points.

When another player thinks you made a mistake, that person may challenge your answer before the next player rolls the dice. If your answer was wrong, the challenger takes the points you would have won, and you score zero. But if your calculation is correct, you get one bonus point for having withstood the challenge.

Play until each player has had ten turns, or five turns each for more than three players. Whoever has the highest total score wins the game.

Keep Score:

Contig Game Board

1	2	3	4	5	6	7	8
28	29	30	31	32	33	34	9
27	55	60	64	66	72	35	10
26	54	125	144	150	75	36	11
25	50	120	216	180	80	37	12
24	48	108	100	96	90	38	13
23	45	44	42	41	40	39	14
22	21	20	19	18	17	16	15

Math Translation

Find an equation in your math textbook, or make up one of your own. Translate it into words with no symbols at all.

For example, here is a famous math translation:

$$\frac{12 + 144 + 20 + 3 \cdot \sqrt{4}}{7} + (5 \times 11) = 9^2 + 0$$

A dozen, a gross, and a score
Plus three times the square root of four
Divided by seven
Plus five times eleven
Is nine squared and not a bit more.

—Leigh Mercer

(Your translation does not have to rhyme.)

Challenge: Read your translation to a friend. Can they write the original equation from your description?

Array Puzzle

Superheroes are lining up in a rectangular array for a parade (or a battle). If they make rows of three, one hero gets left out. If they make rows of five, one hero gets left out. But rows of four work fine. How many superheroes might there be?

Make up your own lining-up-in-rows puzzle.

Career Math

What career are you considering? Write about the ways you will use math in that job. If you're not sure, find someone with that career and interview them.

The Answer Is...

The answer is _____. [Choose any number. Or choose a math vocabulary word.] The question could be ... What?

Can you think of more than one question? How many possible questions can you find for this answer?

Circle Pattern

For this project, you need a compass. Draw a circle about 3–5 cm wide.

Draw another circle the same size, with its center at any point on the first circle's circumference.

At each point where the circles meet, draw another circle of the same size centered there. Continue drawing circles with their centers at each new intersection.

What do you see in this pattern? Can you think of any questions to ask?

This delightful, never-ending circle pattern can be the starting point for many math art projects. Try connecting the intersection points with straight lines to make other shapes and designs.

Number Sums

Write the counting numbers on a single line, with a bit of space between them. On the next line, write another row, so that each number is the sum of the two numbers above it. Keep writing new rows of number sums.

What patterns do you see in the numbers and their sums? What happens if you start with a different set of numbers? What other questions can you ask?

In the News

Read a news article. Write about how it connects to math. Are there parts of the story that could be counted, measured, or graphed? Are there data that indicate a trend?

Examples: slowrevealgraphs.com and nytimes.com/column/whats-going-on-in-this-graph.

Words Help Us Think

Make a math "word wall" on your journaling page. Write all the math vocabulary words you know. Decorate them with frames, or write them crossword-style or in different directions. Or make it fancy any way you like.

OPTIONAL: Include definitions or pictures explaining the hardest-to-remember words.

Chicken Nuggets

At the Math Department Cafe, you can buy chicken nuggets in boxes of 9 or 20. You are eating with a group of friends, and you want to share a total of 38 nuggets. Can you get exactly that number?

What other numbers of nuggets can you buy? Which numbers can't you get? Could you buy exactly 50 nuggets? How about 312? Is there a largest impossible number of Math Department Cafe nuggets?

Litton Game

(two players)

Draw a square array of squares, such as the 4 × 4 array shown. On your turn, shade in as many squares as you like, but they must all be in the same row or in the same column. Whoever colors the last square wins.

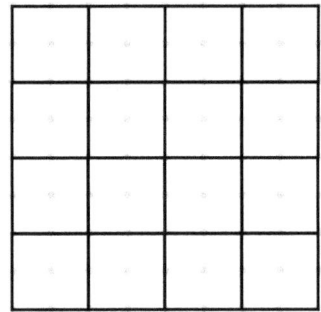

EXTENSION: Mathematicians love to tweak games just to see what happens. How would you modify this game? Test out your new rules with a friend.

Fraction Wall

The top rectangle represents one whole unit. Divide the second rectangle exactly in half. How can you measure where to draw the dividing line? Below that, divide each rectangle exactly into thirds, fourths, etc.

What do you notice? Which fractions are the easiest/hardest to measure?
Color your fraction wall and label the parts.

Visual Patterns

Go to visualpatterns.org and find a puzzle you like. Make a sketch of it in your journal. Describe how you see the pattern growing. Use arrows or diagrams as needed.

   Pick a number from 50 to 100. Can you tell how many would be in that step of the pattern?

Create Your Own Pattern

Draw a pattern that grows according to some rule. Show the first three or four stages of your pattern's life. Can you describe the growing rule with math?

It's Easy

Finish the prompt sentence. And then keep writing until you run out of room. Don't overthink it, just write. Keep your pencil moving. If you can't think of what to write, copy your previous sentence over and over until your mind comes up with something new to say.

The easiest mistake to make in math is...

Infinite Series

Look at the math-art design. What can you say about the shapes or angles? Make a list of the things you notice. What do you wonder? Color the picture, or fill each section with a pattern.

OPTIONAL: Create a related math-art design of your own.

Math Poetry: Limerick

A *limerick* is a five-line poem, usually comical and sometimes quite rude. Limericks use an AABBA rhyme pattern, with three stressed beats on the A lines and two on the shorter B lines.

The rhythm sounds like this:

> da-DUM da-da-DUM da-da-DUM
> da-DUM da-da-DUM da-da-DUM
> da-DUM da-da-DUM
> da-DUM da-da-DUM
> da-DUM da-da-DUM da-da-DUM.

Write a limerick that includes math. For example, here is a limerick about the Collatz conjecture (see also the math equation on page 114):

> A crazy old man, just for fun,
> Liked to triple odd numbers, plus one.
> "Cut the evens in half,"
> He said with a laugh,
> Bouncing up, down, up, down, down, down, done.

Sam Vandervelde's Criss-Cross

(two players)

Draw 3–8 dots spaced out like the corners of a large polygon. Add up to 7 more dots inside the polygon.

On your turn, draw a line (or curve or squiggle) between two currently unconnected dots. If two dots are already connected, you may not draw another line between them. Lines may not cross. The player who can't move loses.

EXTENSION: For more information on the game, including teacher's notes and discussion questions, see the Math Teachers' Circle website: mathcircles.org/activity/game-of-criss-cross.

Silly Definitions

Flip ahead in your math book. Find three vocabulary words you don't know—or try these words: *isometry*, *lemma*, and *equinumerous*. Think up serious or wacky definitions for them. If you wrote the math dictionary, what would these words mean?

Counting Squares

Outline a square array of dots, at least 4 × 4 and up to as large as you wish. How many different sizes of squares can you make by connecting four of those dots using horizontal and vertical lines?

How many squares in all? How does the number of squares grow as your array gets bigger?

What other questions can you ask?

Explain a Mistake

Describe a mistake you made in math, or a problem you missed on a quiz or test. What went wrong? How will you avoid this error the next time? Do you understand the problem now, or is there something more you need to learn about it?

Square Numbers 2

Can you figure out how to draw a square that's tilted on the grid? How do you know for sure it's a square?

Draw several tilted squares of different sizes. How can you find their areas, when the sides don't line up with the grid?

If you draw a tilted square on a grid, the length of its side (S) is usually not a counting number. But that square's area (A) is still the square of the side length:

$$A = S^2$$

And the side is the square root of the area:

$$S = \sqrt{A}$$

Count out the area of your tilted squares, if you can: the whole square units plus the smaller fractional parts that go together to make units. Label each square's area (the square number) and its side lengths (the root).

Living the Dream

If you won $1 million, what would you do with it? Would you have to pay taxes on it? What would you buy? Would you save any of the money, or use it for college, or give some away?

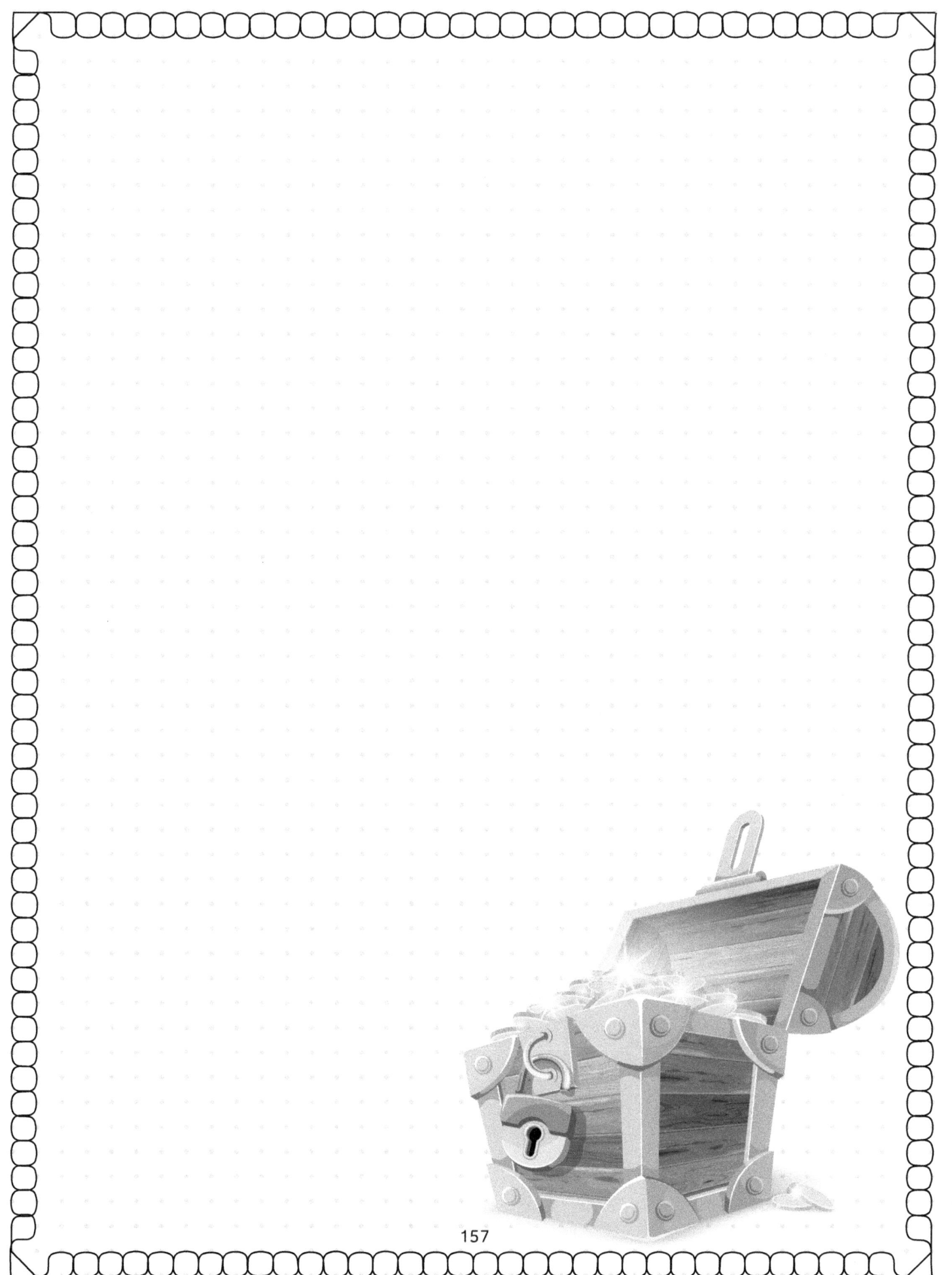

Peter Harrison's Cow Puzzles

The farmer keeps cows in different fields. His fences are labeled with information like the sum of the cows on both sides, or how many more cows are in one field than the other.

Try these puzzles, and then make some of your own.

These fences tell the sum of cows in the fields on both sides:

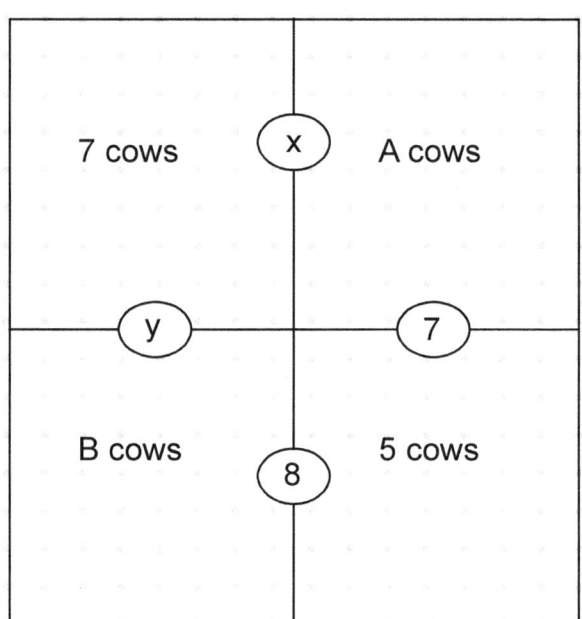

These fences tell the differences between the fields:

That's Mean

The *mean* (average) of four numbers is _____. [Choose any value.]
What might the numbers be? What else might they be?

CHALLENGE: For each set of numbers, also find the median, mode, and range.

Explain How

Think of something you have learned to do in math. Write an explanation simple enough that a younger sibling, cousin, or friend could understand the idea. Add pictures, charts, or diagrams if they help to make your meaning clear.

Mental Math Workout

Practice your Math Rebel skills. Pick any number, and see how many different ways you can write it. Start with a simple expression like addition or multiplication. Write an equal sign and another expression. Think about ways you can modify each expression to create a new one, and keep going until you fill the page or run out of ideas. What kind of fancy math will you create?

☐ =

Secret Number Codes

(two players or two teams)

Each player or team chooses four secret numbers less than 20.

A = _____, B = _____, C = _____, D = _____.

Take turns asking for algebra clues like "What is A + B + C?" or "What is D × A?"

But don't ask "What is A + 3?" or "What is ½ of C?" Players may decline to answer any question that directly gives away one of their numbers.

The first to guess the other player's code wins. Or just play until you solve both codes.

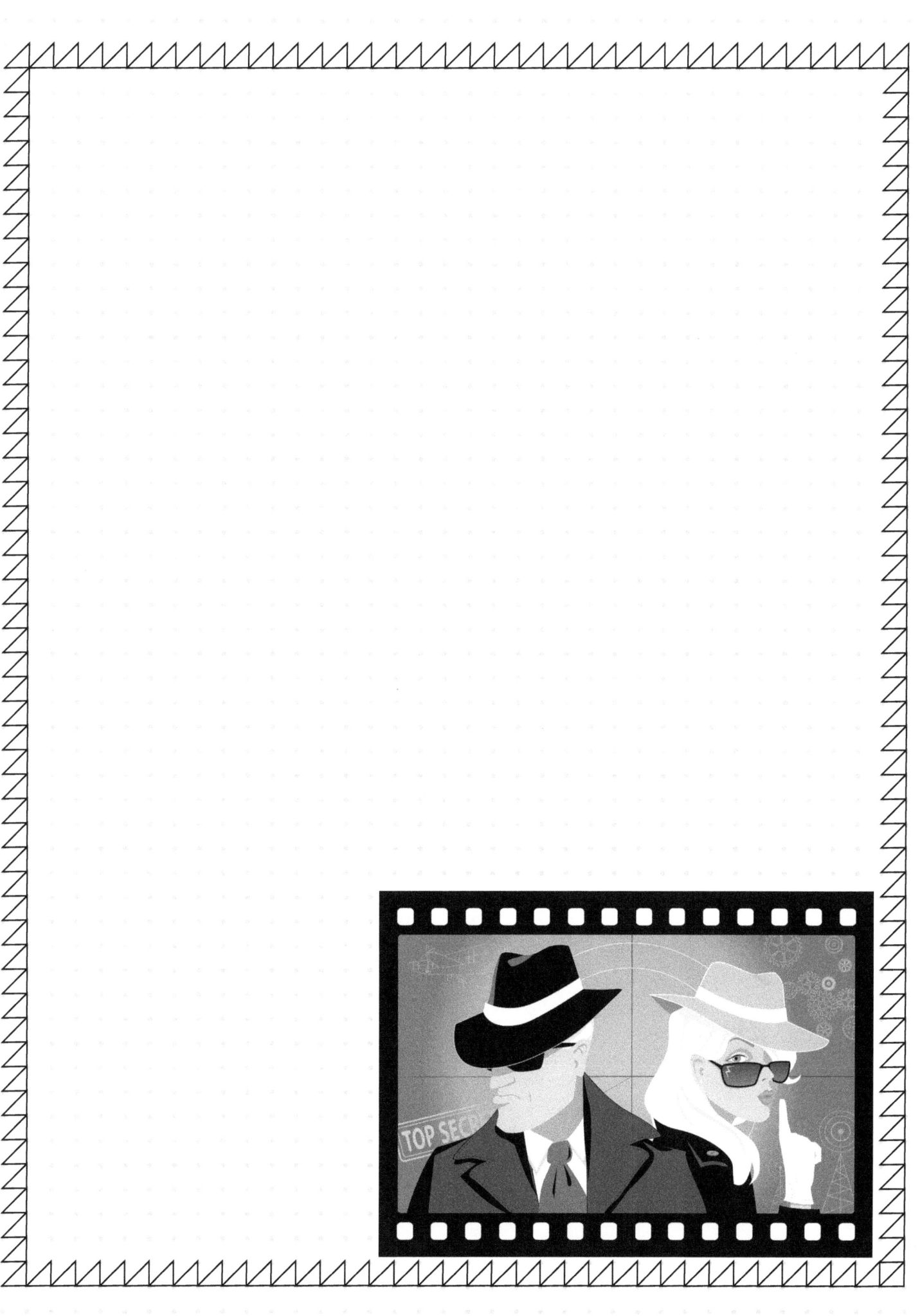

The Mighty Cats

In an ancient Egyptian math puzzle, a rich man's estate contained 7 houses. Every house had 7 cats. Each cat killed 7 mice, which would each have eaten 7 heads of wheat. Every head of wheat, when planted, could produce 7 hekat measures of grain. How much grain did the mighty cats save?

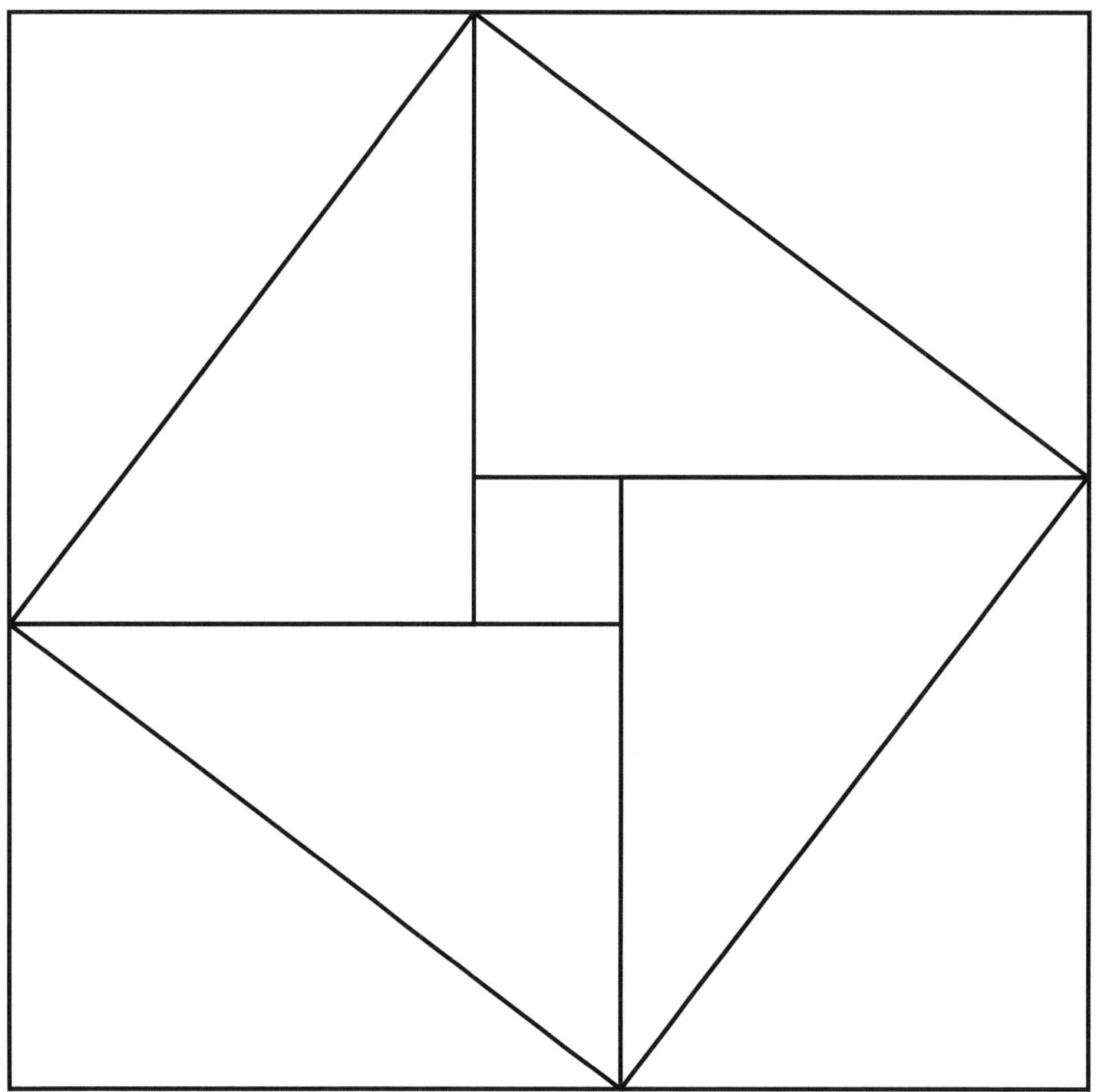

Ancient Chinese Pinwheel

Look at the math-art design. What can you say about the shapes or angles? Make a list of the things you notice. What do you wonder?

Color the picture, or fill each section with a pattern.

[Based on an image in the *Zhoubi Suanjing*, a Chinese mathematical text, date uncertain.]

OPTIONAL: Create a related math-art design of your own.

Don Steward's Swap

Find a set of numbers A, B, and C that will fit this "swap" equation:

$$A \times (B + C) = A + (B \times C)$$

How many sets can you find?

Six-Word Stories

Summarize a math concept or lesson topic in six words. Write it like a riddle:

"Three lines. Three angles. Closed shape."

Or name the topic in your story:

"Three lines. Three angles. One triangle."

One Won Game

(two players or small group)

The first player writes down a starting number greater than 10. Then each player in turn has two choices. You may subtract one from the current number. Or you may divide the current number in half, ignoring any remainder.

The player who reaches the number one wins the game.

EXTENSION: Mathematicians love to tweak games just to see what happens. How would you modify this game? Test out your new rules with a friend.

Math Pickle

Go to mathpickle.com and click the link for "Puzzles," then choose your grade level—or just browse the website. Find a puzzle you like. Try it several times. Write about your experience, what you noticed and what you wondered about.

Triangular Numbers

You've heard of square numbers. *Triangular numbers* are their smaller cousins.

Arrange dots in a bowling-pin pattern. Count the dots to find the triangular number: one dot in the first row ($T_1 = 1$), two in the second row ($T_2 = 1 + 2$), three in the third row ($T_3 = 1 + 2 + 3$), etc.

Keep adding more dots, making each row longer than the previous row. How many triangular numbers can you find?

Do you see any patterns? Can you think of any questions to ask?

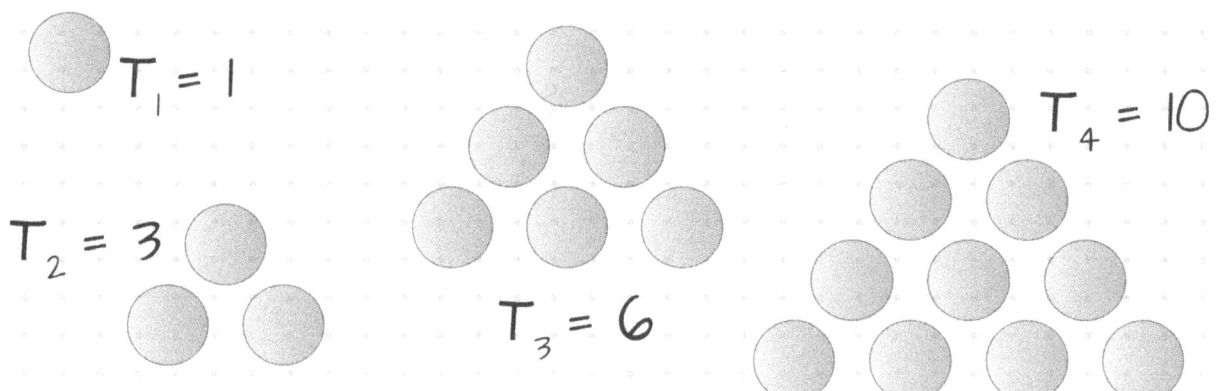

Memory

Finish the prompt sentence. And then keep writing until you run out of room. Don't overthink it, just write. Keep your pencil moving. If you can't think of what to write, copy your previous sentence over and over until your mind comes up with something new to say.

In math, it's always important to remember...

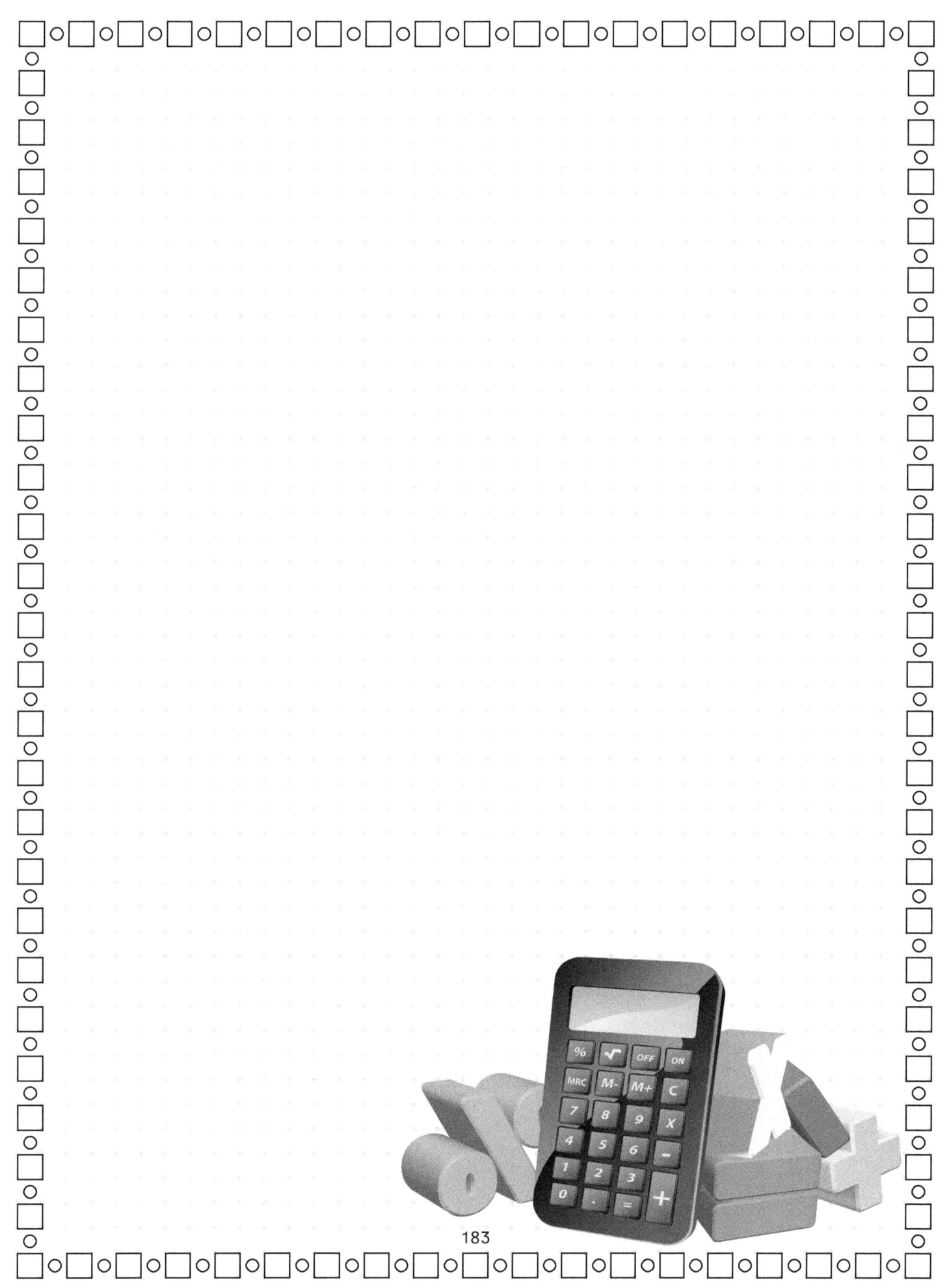

Create a Font

A *font* is a set of specially designed letters, numbers, and punctuation for use in a book, magazine, word-processing program, webpage, or anywhere else people may read stuff. Some fonts are very geometric and regular. Other fonts vary like freeform handwriting.

   Make up your own font. You may want to use graph paper squares to keep your letters consistent. What math do you see in your font? Did you use parallel or perpendicular lines? Are there shapes that repeat in different letters?

   Write your name or some other message in your font.

ABCDEFGHIJKLMNOP

Math Report

Read a math article. Look up something on Mathigon.org or Nrich.maths.org or MathsIsFun.com. Or search for the old Math Munch blog or Martin Gardner's classic Scientific American columns. Write about what you learned. What questions can you ask?

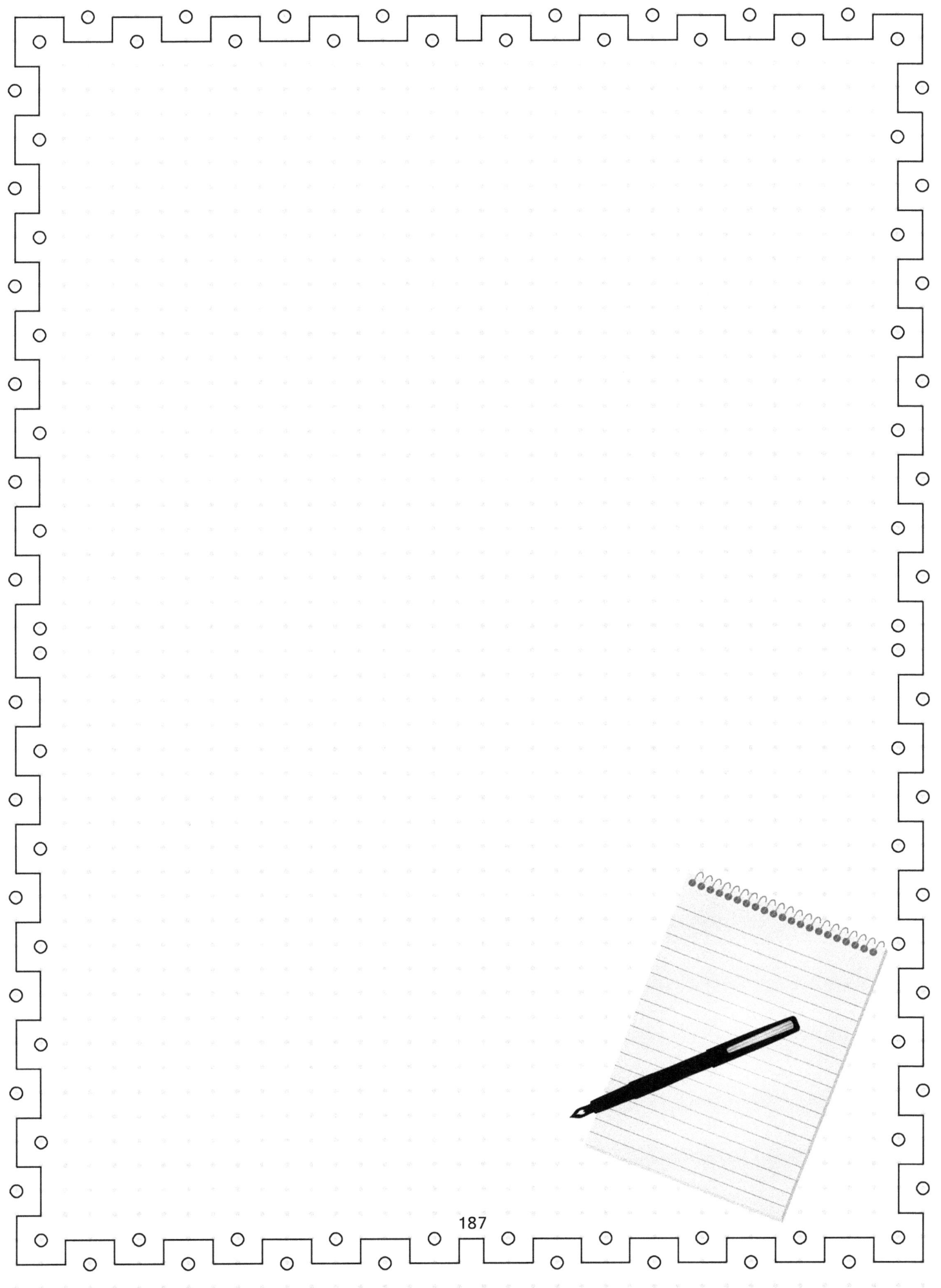

Math Poetry: The Square

Write a poem in which every line has the same number of words as the entire poem has lines. Try to use sensory details and vivid verbs. Your poem does not have to rhyme, but it can if you wish.

For example, you could write six lines with six words in each line. Like this:

Exponential Adventure

Once when famine struck the land—
No rain, no hope of harvest—
The youngest princess launched her quest.
She sought the fabled magic chessboard
That double, double, double, doubles rice.
Enough for all her starving people.

CHALLENGE: Try a longer cubic poem, with the same number of stanzas as it has lines per stanza and words per line. Or a hypercube epic with sections, stanzas per section, lines per stanza, and words per line.

All Ones

Use only the digit 1, and try to use as few of them as you can for each calculation. (Don't just write 1 + 1 + 1 + ... + 1 for every number.) You may use any math operations you know. Can you calculate all the numbers from 1–20? What other numbers can you make?

Place Value Mastermind

(two or more players)

Choose a 3–6 digit secret number. Write your number down, and tell the other players how many digits it has. Each player in turn will guess a number with that many digits.

Write the guess below your secret number, and then tell the others how many digits are exactly correct (but not which ones!) and how many are almost correct (the digit is in your number but not in the right place).

Whoever correctly names your secret number gets to be the next Mastermind, choosing a number for you (and the others) to guess.

Guesses	Number of Correct Digits	How Many Almost Correct
My Number:		

Place Value Mastermind

Keep track of the clues when you are trying to guess another player's secret number:

Guesses	Number of Correct Digits	How Many Almost Correct

Math Riddles Redux

(any number of players)

Choose a secret number the other players will try to guess. Write a "What Number Am I?" riddle. Give at least three clues for your mystery number. No other number should match all the clues.

?

Debate with George Pólya

Decide whether you agree or disagree with the quotation below—and then argue the other side. If you agree, explain why someone might disagree. Or if you think the quote is wrong, tell how someone might argue that it's true.
  [You may also present your own point of view, if you like.]

> It is better to solve one problem five different ways,
> than to solve five problems one way.
>
> —George Pólya

Reinvent Your Homework 2

Find a page of calculations in your math book or download a worksheet online. Answer each question Math Rebel style: Write any true statement *except* what the answer key expects. Have fun making crazy math.

Captain's Log

Keep a math log for a week. Write down all the math you do each day, not just schoolwork. Try to write at least one sentence every day.

Did you learn anything new? Did anything surprise you?

Classify the Aliens

While exploring the alien world of Korb, you meet a variety of local beings from different planetary ecosystems. Your job is to classify the species of Korbian creatures. How will you sort and organize them?

Slime Trail Game

(two players)

Outline any rectangular grid of squares, and mark two "goal" squares in opposite corners, each owned by one of the players.

One player colors any square as the starting Slime, and the other player chooses who will play first.

Players take turns extending the Slime by coloring an empty square horizontally or vertically adjacent to the most recently added square.

If the slime trail reaches a player's goal (irrespective of who marked it), that player wins the game.

Strategic Thinking

Play any math or logic game. Describe your strategy for winning. How do you plan your moves? How do you figure out responses to your opponent's moves?

Don Steward's Arctan Puzzles

For each pair of triangles, tell which angle is greater, α or β. Can you tell without using a protractor? How do you know? Make some arctan puzzles of your own.

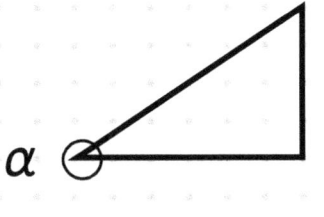

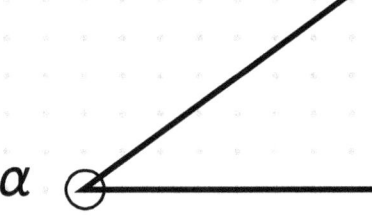

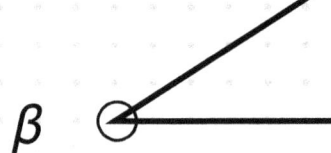

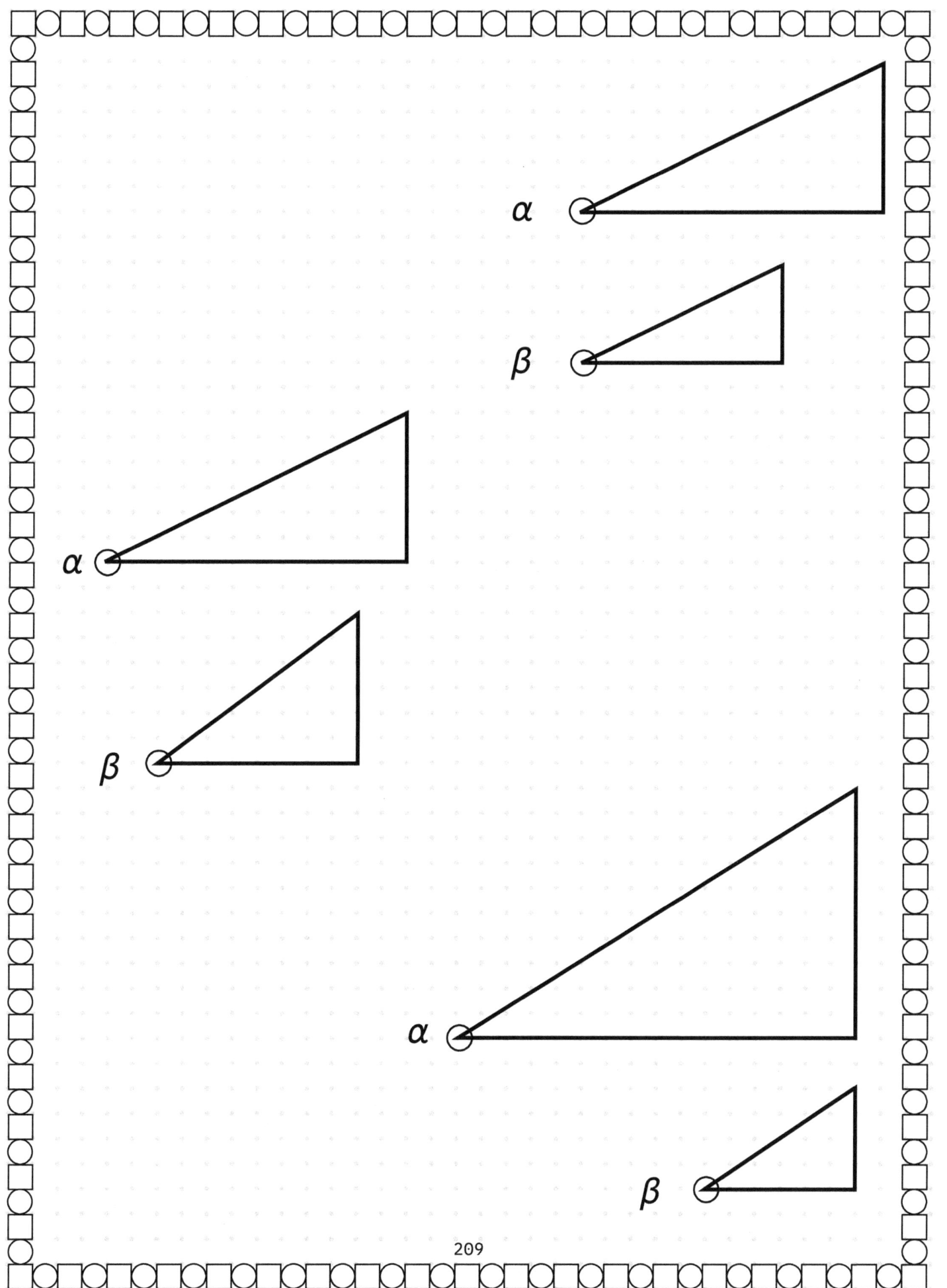

The Substitution Game: Round 2

(solitaire or small group)

Begin with a complexified equation from an earlier round of the Substitution Game. On your turn, copy the equation from the line above, except replace any short expression with an equivalent single number. For example:

$$2 + 8 - 3 = (10^2 - 86) \div 2$$

$$2 + 8 - 3 = (100 - 86) \div 2$$

$$10 - 3 = (100 - 86) \div 2$$

If you end with a true expression like "2 = 2," you win. Congratulations!

If you end with a false expression like "2 = 7," then at least one person made a mistake somewhere in either round of the game.

Don't worry about trying to find the mistake. You still win, because you got to play with a lot of wild and crazy math. So accept the reminder that you're human, and enjoy a laugh at your silly result.

Rear-View Mirror

Has your mathematical imagination grown as you explored the activities and questions in this journal? Explain how, or discuss why not.

If you were the next Math Team leader, how would you guide your students on a creative mathematical adventure?

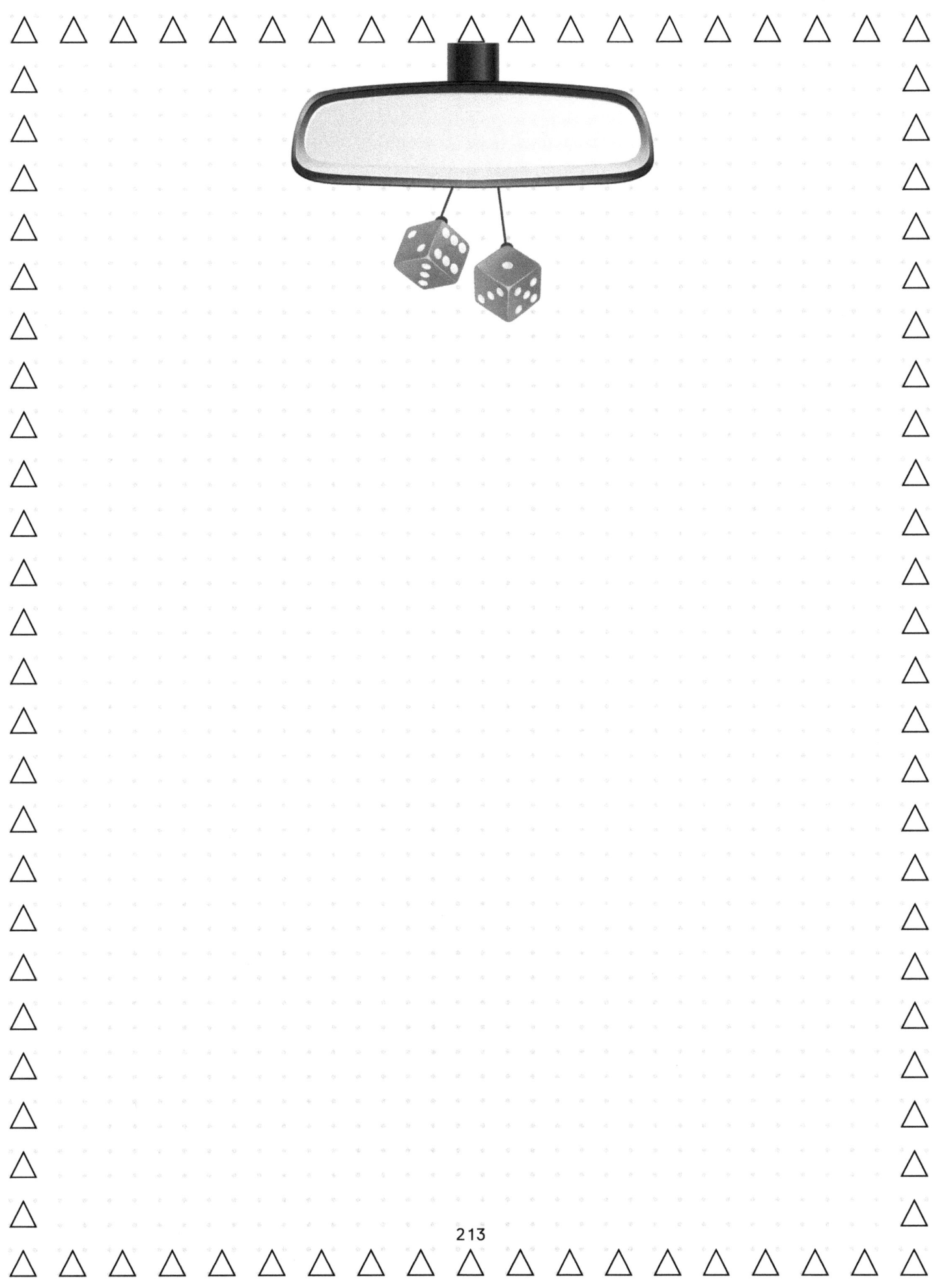

Special Thanks

This book came to production with the help of many wonderful people who backed the project on Kickstarter. Your generous support and encouragement keeps me going!

I'm especially grateful to:

- Abigail, Eleanor, and George Thompson
- Ada Brinker
- Adele & Davide
- Adeline & Ethan Guilford
- Adnane
- Alexandra Stevens
- Alina and Alice
- Angela Z.
- Annabelle, Mary Grace, & Evelyn Stephens
- Ariana Anderson
- Bair Family
- Biswas Family
- Debbie, Charli & Maddox Fear
- Donnell Family
- Edgewood Academy
- Elizabeth Do
- Elizabeth N.
- Emelie Thomas
- Gilles Durand
- Jo O
- Jolie, Dimitri, & Dominic
- Jonathan, Noah, Faith, Joy, Jeremiah, Nathaniel, Grace, and Hope
- Josiah, Alycia, Makelle, Jaxon, Eden, & Asher
- Kayleigh & Aidan Cash
- Lanie Toth
- Lily, Juliana, Arizona, Peyton, Seth, Harper, & Hannah
- Lindy Rose Capel
- Lucas Dietrich
- Lynne Menechella
- Malcolm and Calvin Faull
- MatematicasVedicas.org
- Matilda & Willow
- Matthew, Elizabeth, Mark, Evelyn, and Morgan Johnson
- Moreno Family
- Moriah, Tobias, Abram and Rosalind
- Natural Math Alliance
- Pattie Perry
- Reynolds Family
- Stroman Family
- The Abraham Clan
- The Bancks Family
- The Beaumonts
- The Berman Family
- The Chesebro Family
- The Childers family
- The Cook Family
- The Mickle Family
- The Moore Family
- The Powell Family
- The Prentice Pack
- The Santos Family
- The Whitsell Family
- Thomas Family
- Tracy Popey
- V.G.K.
- Villarreal Family
- Wells Family
- Zëiss, Tesla, Hans, & Luna Knowlton

thank you — grazie — gracias — merci — efharisto — spasiba — danke — arigato — cheers — hvala — obrigado — tack — nkosi — dziekuje — shukria — merkzi — köszönöm — stuutiyi — kiitos — takk — manana — meharbani — matondi — modupe — menlau — sobodi — ngiyabongashukran — buznyg — matondo — grassie — gratzias — shakkran — tanmirt — barkal — nuhun — waybale — miigwech — sulpay — danki — nandri — manjuthe — ahsante — tanemirt — zikomo — dankewol — tashakor — chokrane — bedankt — dakujem — mauuru — blagodaram — marataba — dankegon — yekeniele — shukriyaa — madlobt — misaotra — multumesc — dziakuju — waita — obrigada — spas — kinisou — gràcies — dėkoju — trugèrè — bayarlalaa — blagodaria — talofa — sadol — welalin — supas — skee — vinaka — aabar — mahalo — salamat — dankie — omol — aitäh — dekuji — paldies — taiku — mèsi — tänan — saha — rahmat — wado — murakoze — soolong — tenki — akiba

Discover the World of Mathematics

Textbooks make math feel like a ladder to climb, working rung by rung from one topic to the next. But that's an illusion. Learning math is more like taking a meandering nature walk. Students need to wander around the concepts, notice things, wonder about them, and enjoy the journey.

Most people believe the goal of math is to get right answers. So teachers give us a page of math problems to work, then they check our answers and tell us what we got wrong. When math focuses on right answers, it can make anyone feel like a failure.

Instead, math lessons need to focus on reasoning, on how we struggle through the problems and figure things out. And when mathematical thinking is our goal, those right answers will come along as a side-effect, the natural result of making sense of the math.

—Denise Gaskins

Denise Gaskins' Playful Math

Tabletop Academy Press publishes playful math books for parents who want to help their children build the understanding and skills they need to succeed in school and beyond.

Homeschoolers, afterschoolers, unschoolers, and even classroom teachers appreciate our flexible approach that can work alongside any math curriculum.

Visit us today:
TabletopAcademyPress.com

Or browse Denise's blog:
DeniseGaskins.com

www.ingramcontent.com/pod-product-compliance
Lightning Source LLC
Chambersburg PA
CBHW061148070526
44584CB00034B/4453